W9-AAK-649

PINPOINT

ALSO BY GREG MILNER

Perfecting Sound Forever: An Aural History of Recorded Music

Metallica: This Monster Lives (with Joe Berlinger)

PINPOINT

How GPS Is Changing Technology, Culture, and Our Minds

Greg Milner

W. W. NORTON & COMPANY
Independent Publishers Since 1923
New York • London

Printed in the United States of America
First Edition

For information about special discounts for bulk purchases, please contact
W. W. Norton Special Sales at specialsales@wwnorton.com or 800-233-4830

Manufacturing by QuadGraphics Fairfield
Book design by Charlotte Staub
Production manager: Louise Mattarelliano

ISBN 978-0-393-08912-7

W. W. Norton & Company, Inc.
500 Fifth Avenue, New York, N.Y. 10110
www.wwnorton.com

W. W. Norton & Company Ltd.
Castle House, 75/76 Wells Street, London W1T 3QT

1 2 3 4 5 6 7 8 9 0

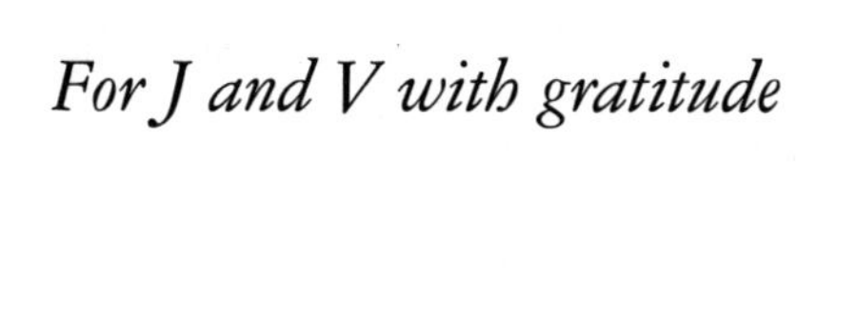

For J and V with gratitude

CONTENTS

". . . consider them both, the sea and the land; and do you not find a strange analogy to something in yourself? For as this appalling ocean surrounds the verdant land, so in the soul of man there lies one insular Tahiti, full of peace and joy, but encompassed by all the horrors of the half known life. God keep thee! Push not off from that isle, thou canst never return!"

—Herman Melville, MOBY-DICK

PINPOINT

INTRODUCTION

The Whisper from Space

Schriever Air Force Base hides in plain sight. A half hour's remove from Colorado Springs, near the western edge of the Eastern Plains, it emerges from the rolling auburn fields visible from lonely Colorado State Highway 94: a few squat buildings and parking lots, interspersed with monstrous domed antennas shaped like mutant golf balls. No aircraft disrupt the serenity of the shortgrass prairie. Schriever is the rare Air Force base that lacks a runway. When he was commander of the 50th Operations Group, Col. John Shaw called his institutional home "the greatest Air Force base the world has never seen."

Schriever's modest footprint contains one of the largest concentrations of classified areas in the service. Many of its 8,000 military and civilian personnel toil underground. Sharing the Colorado Springs area with Peterson Air Force Base, the Air Force Academy, the Army's Fort Carson installation, and the NORAD nuclear bunker carved into Cheyenne Mountain, Schriever accounts for around half of the $6 billion the military pumps into the local economy each year, while maintaining a chronically low profile.

"Where is it exactly?" Shaw wondered rhetorically. "And what do they do there? Who really knows?"

"What we do here is space," said Brian Stewart, a young lieutenant and one of the first people I meet the day I visit Schriever. "We're good at it," he added, "and you'll see why."

With active command over most Air Force satellites, and a lead support role for 175 satellites belonging to other branches of the military, the 50th Space Wing accounts for much of the work done at Schriever. One of its components, the 50th Operations Group, has various units whose duties include overseeing the most top-secret military satellite communications networks. But it is the Group's 2nd Space Operations Squadron whose handiwork resonates the loudest. All day, every day, 2 SOPS has one job: to monitor, maintain, and refine a fantastically complex system that affects nearly every person on earth.

They do their work in a windowless room at the end of a series of fluorescent-lit corridors, past sentries, passcard-protected doors that trigger alarms if left open too long, and signs authorizing the use of deadly force. Inside, around a dozen people, working in twelve-hour shifts, array themselves on tiered rows, staring at monitors. Some have advanced degrees, others are airmen barely out of high school. Overhead screens and rotating siren lights signal the presence of visitors. The captain on duty reminds the crew to hide from view any classified materials. Military police are never far away.

This unassuming space is the Master Control Station for the Global Positioning System. The crews here keep obsessive watch over a constellation of thirty-one GPS satellites, orbiting more than 20,000 kilometers above the planet. Every few minutes, someone at the Master Control Station announces a "pre-pass" and recites steps from a checklist, a prelude to the crew contacting one of the satellites to update its data and perform maintenance. Around the world, from Kwajalein Atoll in the Pacific to the south of England, sixteen far-flung monitoring stations collect data pertaining to each satellite's signal as it comes over the horizon, noting its speed and trajectory. The crunched numbers, based on 1,200 different protocols, tell the crews at Schriever how the satellite is performing, and whether they need to upload new instructions. There is no room for error.

Some people here are monitoring the active atomic clocks aboard each GPS satellite, which are synchronized within nanoseconds to the clocks on every other satellite, all of which obey the Master Clock at the United States Naval Observatory in Washington DC. The satellites broadcast a continuous radio signal that carries information about where the satellite was and will be—and also the exact time the signal left the satellite. The signal makes a 20,000-kilometer journey, taking an especial pummeling as it pushes through the earth's ionosphere. When it reaches us, sixty-seven milliseconds later, it is even fainter. Captain Stephen Dirks, the supervisor of this shift, calls it a "whisper from space."

When somebody fires up a mobile phone in New York, or London, or Tokyo, Karachi, Nairobi, Paris, Berlin, Kabul, São Paulo, Vancouver, Sydney, Budapest, or Johannesburg, the process is the same. Nearly every spot on earth has a line of sight to at least four GPS satellites at all times. The GPS receiver in the phone searches for the four strongest whispers. By noting each signal's origin and its arrival time, the receiver can compute the latitude and longitude of the phone, and express it as a point on a map. The receiver can also provide the correct time. Four satellites, four dimensions. A pinpoint calculation of space and time.*

This extraordinary system began as an American military application, a way to improve the accuracy of bombs and keep

* The GPS receiver performs this calculation by constructing imaginary spheres. The first GPS signal tells the receiver how far it is from the satellite. It could thus be anywhere on the surface of a sphere with the satellite at the center. The second signal creates a second sphere centered on the second satellite, thus placing the receiver somewhere on the circle where those two spheres intersect. The third satellite, by adding a new sphere to the mix, narrows the location down to two points, one of which is obviously wrong—usually miles above the surface of the planet, or deep in the earth's mantle—and can be discarded. The fourth satellite signal resolves any timing ambiguities, since phones don't come with superprecise atomic clocks.

bomber pilots safe. Today, its tentacles are everywhere. GPS is, of course, a wildly popular positioning and navigation system. Nearly 3 billion mobile apps clogging the world's phones and tablets use some sort of GPS-derived positioning information. Between now and 2019, that number will more than double. GPS technology also undergirds an enormous portion of the international economy. The estimated value of the global GPS market in 2011, around $9.1 billion, has now tripled. Certain early GPS entrepreneurs rank among the world's wealthiest individuals. But the true economic influence of GPS resists quantification. Factoring in the GPS chips in smartphones, tablets, and computers, moving platforms, such as cars, ships, and planes, and various products associated with service industries would produce a figure in the trillions of dollars—"so large," according to GPS expert Len Jacobson, "that it is meaningless to anyone but a scholar."

The total number of GPS receivers across all technological platforms probably hovers somewhere around 5 billion. We use GPS to track the movements of criminal suspects, sex offenders, wild animals, dementia sufferers, and wayward children. GPS guides planes to the ground and orients ships at sea. We wear watches with GPS. We buy specialized GPS sporting applications for golfing and fishing. We use GPS to locate oil deposits. GPS has helped grow a significant amount of the food you will eat today.

GPS is itself one of the world's most accurate clocks—and also a clock that unites other clocks. The components and nodes of the world's complex systems require time synchronization, often linked to GPS time. GPS timekeeping helps regulate the electrical grid in all its transnational complexity, bounces your mobile phone conversation from tower to tower, chops up voice transmissions into component parts and reassembles them on the other side, and orders billions of transactions through financial trading networks, where millisecond discrepancies can effect billions of dollars.

GPS can record the movement of subatomic particles across hundreds of miles. GPS helps predict the weather. GPS surveys land, and builds bridges and tunnels. It knows how much water is in the ground and in the ash plume rising from a volcano, and how the oceans help redistribute the planet's center of mass. GPS knows when the earth deforms; it senses the movement of tectonic plates down to less than a millimeter. GPS can help tell us when an earthquake is imminent. GPS can feel the glaciers melting as the planet heats up.

GPS is a global navigational satellite system, or GNSS. Because of the high cost and complex infrastructure of developing and maintaining a GNSS, very few exist. All are controlled by nation-states, and all operate on the same basic technological principles as GPS. Only Russia's GLONASS offers the equivalent full global coverage and a complete satellite constellation. The development of the Galileo system, a project of the European Union and the European Space Agency, did not officially begin until the early twenty-first century. The first Galileo satellite launched in 2011, and the program will require at least another decade to reach full operability. China's Beidou system, currently offering limited service, will mature around the same time.

By then, GPS will be even more fully entrenched. GLONASS, currently plagued by technical problems, will still run a distant second. Many GPS receivers are GLONASS-compatible, using the Russian satellites as a way to strengthen GPS calculations. Even as the European and Chinese systems become more fully formed, and similar projects begun by Japan and India come online, they will likely serve a similar function, support beams in a building whose foundation is GPS. Galileo and Beidou have political value for the countries that run them, a way to declare independence from the United States, but they will likely be "global" only in their coverage, not in their technological ubiquity. The Air Force rightly calls GPS "the world's only global utility." It is universal, free for all, accessible by anyone, influencing everyone. When an ISIS terrorist gets a GPS reading, the process is enabled by the United States military, which presides over every GPS calculation.

The U.S. Department of Defense oversees GPS, with input from the Department of Transportation and other federal agencies, and day-to-day operations delegated to the Air Force. The majority of the sixteen monitoring stations are controlled by the Pentagon's spy agency. Sensors onboard the satellites let them pull double-duty as nuclear-detonation detectors. GPS is an essential part of virtually every weapons system. To maintain it costs more than a billion dollars a year.

Like the Internet, GPS arose partly out of Cold War imperatives. (It is no historical accident that the closest facsimile to GPS today is GLONASS, begun by the Soviets a few years after the US launched GPS.) Although it is less visible, GPS's influence on the world equals or exceeds that of the Internet. (The Internet could not operate without precision timing controlled by GPS.) This seems odd because while the Internet is a vast database, a way to aggregate and share information, GPS is just a radio pulse, a descendant of the rhythmic blip emitted by Sputnik. But this whisper is so dependable, so ordered and clean, that GPS has become our heartbeat. If it failed tomorrow, our society would experience enormous disruptions and scientific setbacks.

Those who oversee GPS have to maintain an almost unimaginable degree of exactitude. Something as subtle as the pressure of the sun's rays can shift a satellite's orbit. And the clocks cannot falter. The integrity of the entire system rests on measuring the distance between you and the satellite by timing the arrival of its pulse. Those signals are traveling at the speed of light. A timing error of just one-millionth of a second will translate into a distance error of 200 miles. Put another way, a clock with a margin of error of .000001 seconds might locate a New Yorker close to Washington, DC, or a Parisian in the vicinity of Brussels. Your GPS receiver, the one in your phone, is accurate to within a few meters.

Even with all that—the Master Control Station engineering the satellites, the clocks measuring time with rubidium atoms, so accurate they will not fall out of lockstep in millions of years—it still isn't enough. Like particle accelerators, and few other human-made operational systems, GPS must account for Einstein's laws of relativity. Compared to clocks on earth, time passes slightly more slowly for the clocks onboard the satellites, which speed around the earth at 2.4 miles per second. The GPS signal contains instructions for the receiver to correct its calculations accordingly.* The difference is just a few microseconds every day, but if the system did not account for it, timing errors would multiply and distance calculations would soon be off by thousands of miles.

This book tells the story of GPS and how it grew from a fledgling military project to a ubiquitous technology that blankets the world. The heartbeat emanating from those thirty-one GPS satellites gives us the power to measure and obtain enormous amounts of information about our planetary environment, physical space, and human behavior. Its invention has led to an explosion of creativity in science, technology, and business. For better or worse, it forms an essential part of the infrastructure of modern life.

But there is a price: the system may fundamentally change us as human beings. We so rely on GPS, have integrated it so deeply into our lives, that it may be altering the nature of human

* The curvature of spacetime caused by the earth's mass makes the satellite clocks appear to run a bit *fast*. So the receivers also correct for that at the same time (such as it is).

cognition—possibly even rearranging the gray matter in our heads. It is so potentially invasive that it forces us to reconsider cherished notions of privacy. We have let it saturate the world's systems so completely that it is difficult to imagine life without it, and so quickly that we are just beginning to confront the possible consequences. A single GPS timing flaw, whether accidental or maliciously installed, could bring down the electrical grid, hijack drones, or halt the world financial system. We now trust our devices so much that we follow them blindly down abandoned roads, over cliffs, and into the ocean; park rangers call this "death by GPS."

The story of GPS is also one of human ingenuity. At Schriever, they tell visitors that GPS is officially divided into three segments. The first is the space segment: the satellites. Next is the control segment: the tracking stations and the facilities that upload ephemeris corrections to the satellites. The third, the largest, is the user segment: every one of the world's GPS receivers. Workers at the Master Control Station like to point out that they are only responsible for the first two. Once the signal leaves the satellite, their job is done. The rest is up to us.

GPS reaches us as a whisper. We give it a voice. Listen closely, and it will tell the story of the world today.

PART ONE

Calculating Route

CHAPTER ONE

Tupaia Goes Home

> *"I was hypnotized by the road. I was leaning forward and I let the speed gradually creep up . . . On the long empty stretches I tried to imagine I was stationary and that the brown earth was being rolled beneath me by the Buick tires. It was a shaky illusion at best and it broke down entirely when I met another car."*
>
> —Charles Portis, THE DOG OF THE SOUTH

The habitable earth was once an island.

Around 250 million years ago, the only land above sea level was a single mass called Pangaea. Over eons, the great continent fractured and the movement of earth's plates produced the geography we know today. But the movement never stopped. Right now, Africa is on a collision course with Europe. (Payback!) Australia is drifting north, taking aim at Eurasia, poised to scoop up the islands of Southeast Asia. A huge underwater ridge that divides the Atlantic Ocean is growing, at about the same rate as your fingernails, increasing the ocean's size and exerting pressure on the Americas. By the time the waters recede, these lands will also be swimming toward Eurasia.

We are living at the midway point of this geo-diaspora. In another 250 million years all the world's land will once again merge into what geologists are calling Pangaea Ultima.

Fifty thousand years ago, Pangaea was a dim geological memory, but the planet did not look quite the same as it does today. Glaciers had caused sea levels to drop by as much as 100 meters in some places. In Southeast Asia, the receding ocean had revealed a peninsula called Sunda that connected Indonesia with Asia—

and a continent called Sahul that comprised New Guinea, Australia, and Tasmania.

The watery distance between Sunda and Sahul was small enough that some adventurous inhabitants of Asia began to explore the offshore lands with relatively simple boats. It was humankind's first tentative migration away from the African/Asian supercontinent. The most daring among them soon ventured out into the wider Pacific Ocean, settling as far away as the Solomon Islands, a few hundred miles east of New Guinea. For many thousands of years, they went no further—and it's not hard to understand why. The ocean must have seemed unfathomably vast—perhaps even more so than the night sky, whose familiar stars suggested some sort of logic or boundary. The Pacific Ocean covers one-third of the globe, an area equivalent to all of the land in the world above sea level. When they stared out at the water, they were seeing a Pangaea in negative.

Around 45,000 years later, a new oceangoing migration began, originating from what is now the Fujian province in China. Over the next 1,500 years, the Austronesians spread to Taiwan, eastern Indonesia, and New Guinea. Eventually, some of them did something even more remarkable. They decided to venture into the big blue void. They sailed canoes across the Ring of Fire, the seismic and volcanic hotspots of the Pacific Rim, out into what we now fittingly call Remote Oceania.

After a few hundred years, the Austronesians had discovered and settled Fiji, Tonga, and Samoa, the ancestral homelands of the region now called Polynesia. The Polynesians learned more about oceangoing, and extended their reach to even remoter Oceania, locating the Cook Islands and the Marquesas. Further out, the Pacific really opens up. Land is even scarcer, but the Polynesians found much of it. The islands of Polynesia—not unified politically, but closely related culturally and linguistically—eventually formed a triangle in the Pacific, with points marked by Rapa Nui (Easter Island) to the east, Aotearoa (New Zealand) to the west, and way up north, Hawai'i, among the most isolated archipelagos on the planet, with no further land for 2,400 miles to the east and more than twice that to the west.

It is difficult to overstate the immensity of this accomplishment. On the grandest scale, the establishment of Polynesia was but part of the larger Austronesian conquest of the planet. Austronesian languages spread west nearly as far as Africa, and east to within a few miles of the Americas. But the establishment of Polynesia was the last great premodern human migration.

In seafaring and navigational terms, while the Europeans were discovering fire, the Polynesians had already split the atom. They crossed the ocean in canoes roughly 60 feet long, built from hollowed-out trees, with sails made from woven leaves. To avoid capsizing from the awesome power of the wind on the sails, the Polynesians made their longest trips in double canoes, joining two hulls to create a catamaran-style vessel with a width of about 20 feet.

The wind came primarily from the east, meaning migrants sailed directly into headwinds, like moving through an atmospheric wall of tar. And yet, with no compass, sextant, or any other modern navigational aid, explorers in canoes found tiny oases scattered across one-third of the planet, an expanse nearly as vast as Europe and Asia combined. For at least a century after this migration was completed, European navigators were still wary of sailing their ships beyond the Strait of Gibraltar into the Atlantic Ocean.

How did they do it? A lot of trial and error, of course, with much human sacrifice. During the two millennia the Polynesian migration was underway, as many as 500,000 souls were lost at sea. In their double-outrigger canoes, they spread across thousands of miles of ocean, through overpowering waves, swells, currents, and storms. Everything in their physical environment was poised to work against them, but they conquered it. This was no random quest. Computer modeling of currents and atmospheric conditions has demonstrated the extreme unlikelihood that explorers discovered islands through uncontrolled drift. Their ability to know where they were and where they were going, across a seemingly faceless ocean, is baffling.

A Polynesian navigator would have several tools at his disposal to set a course for travel. Chief among these was a sidereal compass, a complex ingrained mental image of the stars. Beginning in childhood, the future navigator would spend years honing the ability to gather clues from the environment while out on the ocean. Over the centuries, these methods would be combined with a growing knowledge of the placement of islands, as the migration spread further.

But this young navigator would also have something else, something that eludes simple description—call it a worldview. Its intricacies remain a subject of fascination today, and we are still piecing together its components. The most celebrated European explorer of the Pacific could not understand the map of the world Polynesian navigators had in their heads, even when he had the good fortune to meet someone willing to sketch its contours.

In the early sixteenth century, Ferdinand Magellan completed his historic voyage across the Pacific Ocean. For the next 200 years, European ship captains considered the Pacific to be, at best, a distance to be traversed on the way to somewhere else. That began to change in the mid-eighteenth century. Expeditions became prestige operations, underwritten by governments interested in exploration for its own sake, and convinced that knowledge of the location of remote islands could offer advantages, both commercial and geopolitical. The conception of the ocean as empty space was replaced by something mysterious, unwieldy—a watery Other that mocked attempts to understand it. "The Pacific no longer appeared as it had done to Magellan a desert waste," the scientist Alexander von Humboldt noted from the vantage point of the mid-nineteenth century. Instead, "it was now animated by islands, which, however, for want of exact astronomical observations, appeared to have no fixed position, but floated from place to place over the charts."

By the time James Cook, a British sea captain, embarked on his first voyage of the Pacific, in 1768, fewer than 500 European ships had crossed the ocean since Magellan, and only around two dozen of them had made landfall on any Pacific islands. The first European to sail to the Pacific with the primary goal of exploring it, Cook obsessively charted the islands, creating the earliest detailed map of the Pacific, and he and his crew went ashore whenever they could. Although this contact would eventually have a catastrophic effect on the inhabitants of these islands, Cook obeyed the instructions of the Royal Society, which underwrote his voyages and instructed him to respect the sovereignty of any native peoples he encountered.

When Cook's ship, the *Endeavour*, arrived in Tahiti in 1769, he learned some of the local language, and was struck by its similarities to what he had heard spoken in Aotearoa, 2,000 miles away. Cook eventually concluded that the islands of Polynesia were linked, forming "the most extensive Nation spread over the face of the Earth." That realization led to a vexing question, recorded in his journal in 1778, during his third and final Pacific voyage: "How shall we account for this nation spreading itself so far over the ocean?"

The irony of Cook asking that question was that by then he had already found someone who *could* do exactly that. While in Tahiti, the captain had met a man named Tupaia, who hailed from the nearby island of Ra'iatea. A charismatic figure with a free-ranging intellect, described as "the Machiavelli of eighteenth-century Tahiti" by his biographer, Tupaia had a knack for getting into—and out of—dire trouble. When invaders from Bora Bora had sacked Ra'iatea a few years earlier, he managed to escape to Tahiti with a sacred icon of 'Oro, the god of war. 'Oro worship soon swept Tahiti, and within a few years Tupaia was Tahiti's highest-ranking priest. He later became involved with Puea, the ex-wife of a chief, and the two hatched a disastrous plot to become the island's rulers. A brutal internecine war broke out, and the two fled for the hills.

When Cook arrived, Tupaia, now stripped of power, sought him out. (Cook had heard of Tupaia from another British captain who had reached Tahiti and met him.) Besides the obvious cultural differences, Tupaia and Cook were an odd match. Unlike most British captains, Cook did not come from the upper classes, which had made it difficult for him to rise through the ranks. Tupaia, until his recent troubles, was a member of the *arioi*, an elite society built around 'Oro worship, open to only those deemed exceptionally physically attractive and gifted. Some of the *arioi*, including Tupaia, were trained navigators, and it was this skill that made him particularly interesting to Cook.

Tupaia got to know Cook and his crew throughout the spring of 1769. Hoping to leave his troubles behind, Tupaia asked the captain if he could accompany him on the voyage. The consensus among the captain and the crew, Cook later wrote, was that Tupaia was "a very intelligent person," someone who knew more about "the Geography of the Islands situated in these seas, their produce and the religion laws and customs" than anyone else they had met on the island. Perhaps because of his dispossessed status, Tupaia was willing to share closely guarded navigational skills that no Polynesian had previously taught to a European. On September 9, Cook recorded, the *Endeavour* set sail, "in search of what chance and Tupia might direct us to."

Tupaia was a hit aboard the *Endeavour.* Although skeptical of his ability to conjure strong winds through prayer—Joseph Banks, the ship's naturalist, maintained that Tupaia would begin praying when a breeze was already headed their way—Cook and his crew were amazed by Tupaia's powers of orientation. At any moment, Tupaia could cast his gaze out into the infinite ocean and point in the exact direction of Tahiti.

As Tupaia grew closer to the crew, he began to share his vast knowledge of the region's geography. Working with Cook and the ship's master, Robert Molyneux, he listed some eighty-eight islands he knew. With Banks's help, they drew up a chart of Ra'iatea. As cross-cultural exchanges go, this was a big one, an unprecedented attempt to express this ancient geographic knowledge in the language of Cartesian space. Tupaia also began to describe some navigational methods, including his knowledge of astronomy and, crucially, the way Polynesian sailors had learned to deal with the prevalence of easterly winds by taking advantage of the few months when stronger westerlies blew.

Soon after taking on Tupaia, the *Endeavour* visited Ra'iatea. Walking along the beach during a celebration held in honor of the ship's arrival, Cook and Banks encountered several boathouses containing canoes. The sight triggered an epiphany for Cook. "These people sail in those seas from Island to Island for several hundred Leagues, the Sun serving them for a compass by day and the Moon and stars by night," he wrote. "When this comes to be prov'd we Shall be no longer at a loss to know how the islands lying in those Seas came to be people'd, for if the inhabitants of [Ra'iatea] have been at Islands lying 2 or 300 Leagues to the westward of them it cannot be doubted but that the inhabitants of those western Islands may have been at others as far to westward of them and so we may trace them from Island to Island quite to the East Indias."

Tupaia offered some proof. During his time on the *Endeavour*, he produced a map of the Pacific that showed Tahiti and 130 islands. The map spanned a distance of 2,600 miles—roughly the size of the continental United States—east to the Marquesas and west to Rotuma and Fiji. With the exception of Hawai'i and Aotearoa, outliers in the triangle, every major Polynesian island group was covered. The map not only displayed Tupaia's extensive geographical knowledge, it was also a tacit argument for the power and precision of Polynesian navigation. Tupaia's sophisticated expertise, passed down from generation to generation, argued for the ability of his ancestors to travel great distances, even back to islands settled during earlier periods of the great Polynesian migration.

Given Cook's belief that the Polynesian islands comprised "the most extensive Nation spread over the face of the Earth," and his recent insight that its inhabitants had developed the means of traveling throughout this nation, Tupaia's map was an invaluable resource. Yet there is no evidence that the map had much of an impact on Cook. He mentions the chart in his journal only once: "the above list was taken from a chart of the islands drawn by Tupia's own hands." It was a difficult map to read—the positions of the islands in relation to Tahiti were sometimes puzzling in light of what was known about the geography of the Pacific—but Cook did not ask Tupaia to elaborate. Nor did he ask Tupaia how he had acquired this knowledge.

Tupaia became increasingly alienated from the crew. He urged Cook to point the ship west, promising to show him many islands he'd never seen, but Cook chose to go south, looking for *Terra Australis*, the fabled southern continent. During this period, Tupaia doesn't even appear in the ship's log. He contracted scurvy, and then dysentery, which caused his death during the winter of 1770. Cook's sole tribute in his journal described Tupaia as a "Shrewd, Sensible, Ingenious Man, but proud and obstinate which often made his situation on board both disagreeable to himself and those about him." For the rest of that voyage, and the two others that Cook led in the Pacific, he never solved, to his satisfaction, the puzzle of how Polynesians settled the Pacific.

One day in 1970, the bicentennial year of Tupaia's death, David Lewis received a letter that ended on a cryptic note: *Now I am still alive. But you will meet me one day or not? Because I am getting old.*

Lewis, an Australian physician, adventurer, and master sailor, had won a university fellowship to sail the Pacific in search of anyone who still practiced the ancient art of Polynesian navigation. It was a dispiriting experience, as Lewis discovered that the practices had died out, superseded by modern methods. Then he met Tevake, a man in his seventies who had begun his navigational training when he was seven or eight years old. In his younger years, Tevake would regularly sail a 30-foot outrigger canoe on a journey of 300 miles or more. Age had slowed Tevake somewhat, but he still traveled solo between his atoll and nearby islands.

Tevake was "the first Polynesian navigator I ever sailed with," Lewis remembered, as well as "one of the greatest." This gifted navigator lived on an atoll in the Santa Cruz Islands, part of the Solomon Islands, a country in the area of the Pacific called Melanesia, west of the Polynesian triangle. Tevake's home was not, therefore, part of Polynesia, but Lewis realized this was as close as he would get. For scholars like him who were desperate for information about Polynesian migration, time was running out. Lewis described Tevake as "almost alone, one of the last in a line of navigators that stretched back 5,000 years to the Austronesian dispersal."

Tevake was Lewis's Tupaia, and Lewis was determined to be a better protégé than Cook. "I became his pupil," Lewis would later write. "He became my friend." They even looked a little alike, each with a broad nose and a shock of curly hair bleached by years under the sun on the open water.

Lewis was part of a subculture that had formed in the two centuries since Cook, comprised of people—many of them amateur scholars located around the Pacific Rim—going to great lengths to advance their theories about the mysteries of Polynesian migration. Joaquín Martínez de Zúñiga, a Spanish priest in the Philippines, wrote a book in which he argued that Filipinos were descended from explorers in Chile and Peru sailing westward. John Lang, a Presbyterian minister in Australia, devoted two books to his contention that, yes, the Polynesians had Austronesian origins, but no, they had not meant to sail so far east—caught up in winds blowing that direction, they had drifted to the various islands, and all the way to South America.

In the latter part of the eighteenth century, a new school of thought emerged that reached the right conclusions for the wrong reasons. Abraham Forlander, a newspaper editor and judge living in Hawai'i, and S. Percy Smith, a New Zealand land surveyor who founded the *Journal of the Polynesian Society*, found evidence for eastward migration in local legends and oral traditions. These theories had a veneer of scientific rigor, but they were undercut by a persistent exoticized treatment of the Polynesian migrators as almost superhuman in their ability to navigate the Pacific. This supposedly positive spin on the Eurocentric idea of the "native" as unknowable Other precluded any rational explanation of how they had accomplished this feat.

As late as the mid-twentieth century, there were those who could not accept the mounting ethnographic and archaeological evidence that Cook's theory of eastward migration was correct. In 1947, Thor Heyerdahl, a Norwegian adventurer, made headlines worldwide by maneuvering a balsa wood raft called *Kon-Tiki* from Peru to the Tuamotu Archipelago. He argued, in a subsequent book, that his voyage proved that the Polynesians' origins were in South America. Andrew Sharp, a retired New Zealand civil servant, believed the Polynesians' ancestors were Southeast Asian, but found laughable the idea that their migration was controlled. "Most people believe what they want to believe, and most people want to believe that the Polynesians sailed back and forth to their distant lands without quadrant, compass, or chart," he wrote. "The view that they were supermen is not satisfactory basis for a theory of Polynesian long navigation." Rather than evidence of navigational prowess, Sharp believed Tupaia's map reflected the accumulated geographic knowledge of people who had drifted to Tahiti, compiled and synthesized by the very clever Tupaia.

In the postcolonial climate of the 1960s, Lewis and a few other like minds wanted to impart human agency to these seemingly mythic navigators of yore, to ground their migratory achievements in actual methodology. Lewis felt he had the seafaring knowledge—and, in Tevake, an invaluable resource—to discover the inner workings of Polynesian navigation that made the migration possible. During his time with Tevake, Lewis learned how the older man was taught to memorize star positions, and how he gleaned information from careful observations of currents, swells, and wind. But even as they grew closer, Lewis felt there was something he was missing. It was as though he couldn't quite manage to see the world through Tevake's eyes. So Lewis felt great dread as he read Tevake's mysterious missive. What did that mean—*you will meet me one day or not*?

Some months later, Lewis found out, in a letter he received from the Santa Cruz District Office. It explained that Tevake had recently said goodbye to his neighbors and the next day embarked on a journey to a nearby island—a voyage, the officer noted, that "would be nothing for a seaman of his caliber." He never reached the island, and was never heard from again. After his disappearance, when his neighbors considered how Tevake had said goodbye, they agreed that his manner was oddly formal, as though he had made up his mind to end his life in the waters he knew so well. Some speculated that he had experienced a premonition of impending doom. Or perhaps he had just decided it was time to go. Whatever happened, Lewis felt the world had lost more than just an expert navigator. Tevake represented the last link to firsthand knowledge of how humans had conquered the most treacherous third of the world. And Lewis had failed to extract that knowledge before the link was severed.

Lewis looked again at Tupaia's map. He suspected it represented a way of conceiving space that was similar to Tevake's—literally, a completely different worldview.

Some principles of ancient Polynesian navigation are incredible to behold, but conceptually simple to grasp—that is, we can *imagine* how they would work. We understand that any navigator must first set a course. The Polynesian navigator's primary tool would be his sidereal compass. Not a compass in the way we understand the term, the sidereal compass cannot be held in the hand—it is all in his head. It is the memorized knowledge, learned over many years of training, of the positions of stars and how they relate to islands in the navigator's region. The navigator plans a route using various star-points to mark the course.

He has ways to augment his navigating, especially when stars aren't visible due to daylight or weather. The sun provides an east–west heading at sunrise and sunset, and defines north at noon. Long-distance wave systems called swells are fairly consistent and predictable—the navigator knows the various (often overlapping) swell patterns in his region of the Pacific and the directions they are traveling, knowledge that helps to orient the canoe. At sea, he might pause to "read" the swells hitting the canoe, while editing out the "noise" of more volatile local waves. The navigator might even squat in the canoe, dragging his testicles on the floor to feel the vibrations.

Wind and currents constantly threaten to make the canoe drift off its prescribed course. To make the necessary corrections, the navigator needs to know how his specific vessel reacts to wind, and how currents vary by region, season, and the ocean floor's topography. This is all happening constantly, through extremely powerful currents and wind that is rarely directly at his back, blowing in the direction he wants to go—so he is always calculating and recalculating, tweaking the canoe's course to counteract these forces. At times, the canoe will be caught in a gale so powerful that all the navigator can do is ride it out, keeping track of where the storm is taking him, and resume his course when it passes.

If the navigator maneuvers through all these obstacles, and senses that the canoe is nearing land, he looks for helpful signs. He scans the skies during the day, searching for certain bird species and noting cloud formations. At night, he looks for the luminescent flashes caused by microorganisms deep below the surface, aware that they usually point in the direction of nearby land.

By the end of his journey, the navigator has drawn on concepts relating to astronomy, oceanography, meteorology, geology, ornithology, physics, and marine biology. It is difficult for most of us to imagine setting a course using only memorized knowledge of the stars, let alone exerting the mental energy to maintain it on the punishing Pacific. But we get the concept, because everything the navigator has done involves holding steady to a course, maintaining a bearing, keeping the canoe heading in the direction he wants to go.

And yet, if this was the extent of his skills, he and everyone else on the canoe would surely be swallowed by the waves. There is something else he needs to know, more data he requires. Grasping what he does to get it is the point where our comprehension falters, our fingernail hold failing as we slide off the canoe of the mind into a dark Pacific of the soul.

If I live in a house that lies just off a rural highway, hidden by a grove of trees, accessible by an unmarked, unnamed dirt road that appears on no maps, and I'm giving you directions to reach it from your house, in a city south of me, I can't just tell you to take the highway north. That would be correct, but useless. I have "set a course" for you—you will stay on the ribbon of pavement that is the highway, heading north, maintaining your course by not allowing your vehicle to drift off the asphalt—but you won't have any idea when to start looking for that turnoff. Almost as uselessly, I could tell you I live north of a certain gas station. Again, true as far as it goes, but so what?

I would need to tell you that my house is located about 10 miles north of that gas station. This would orient you. You could reset your odometer at the gas station and count the miles. My directions, in essence, require you to know where you are at all times—information the odometer gives you. You can look at it and say, "At this moment I am seven miles from my origin and three miles from my destination." The odometer allows you to perform a very simple variation of the type of calculations that navigators call dead reckoning: defining one's current *unknown* location by referencing a previous *known* location, using such information as the time elapsed since that point, the average rate of travel, and the course heading. By counting wheel rotations, the odometer provides a way to conceptualize your location, to understand what "here" means.

The Polynesian navigator has set a course defined by a series of star-points. To execute the course, to know when to stop following one star and start following another, he needs a way to define his location at all times, which requires a highly sophisticated form of dead reckoning. His training has made him adept at gauging the average speed of the canoe, through close observation of wind, turbulence, and the spray of water in the canoe's wake. Close observation of the movement of the sun and stars allows him to gauge the passage of time. The ability to solve for those two variables—the rate of travel and the elapsed time—makes the Polynesian navigator a powerful dead reckoner. But he still needs a way to visualize where he is and how much he has progressed on his journey, in the same way the odometer's calculations give you a sense of how close you are to my house.

One reason it is so difficult to understand the Polynesian navigator's method of pinpointing himself is that we are no longer really talking about navigation. Navigation is the process of determining a route. The way we visualize executing that route is called wayfinding. As behavioral geographer Reginald Golledge put it, wayfinding is the way we "embed the route to be taken in some larger reference frame." A full understanding of Polynesian navigation requires us to poke our heads through a completely new frame.

Imagine the earliest human society—not so much a village as an encampment. For its inhabitants, this encampment defines their world. Whenever they leave it to hunt or gather, they conceive of their present position in relation to their encampment. They know no other way to think of location. This is their reference frame.

The great aviator Harry Gatty (his records included the fastest circumnavigation of the planet, in 1931) used the term "home-center" to describe this basic wayfinding method. He contrasted home-center systems with what he called self-centering methods, which is how we see the world. We define our position objectively, sometimes with help from tools such as maps or GPS, but rarely by where we are in relation to our homes. Dead reckoning is kept to a minimum. The center of our world is us.

Gatty, who had some knowledge of ancient Polynesian navigation, identified a third system, which he called "local-reference." This defines location in relation to a prominent environmental feature, such as a mountain, coastline, or bridge. Newcomers emerging from the New York City subway into the disconcerting tangle of the West Village—where 4th Street and 13th Street somehow intersect—can use the massive One World Trade Center building to distinguish uptown from downtown.

Lewis suspected Tevake had used a local-reference system that had vanished from Polynesia but was still practiced in the Caroline Islands, a method of dead reckoning called *etak*. The first key to understanding *etak* is that it does not provide the navigator with any new navigational knowledge. *Etak* is wayfinding, not navigation. It will not tell you how to get to where you want to go, any more than knowing what "miles" are will help you get to my house. But it provides an even more elusive, more primal, more contingent form of knowledge: it will tell you where you are.

This is how it works:

When a navigator using *etak* decides to travel from one island to another, a third island is chosen as a reference point. The reference island will be one that lies between the two, and off to one side. (Picture a triangle.) The reference island *is not visible from the origin or the destination, and will never once be visible during the journey*. But the navigator knows, from his sidereal compass, the bearing of the reference island. He knows that when he stands on his island and looks in the direction of a certain star, he is looking in the direction of the reference island.

He begins the journey. This one will require three segments, or *etaks*. (Most journeys required several more.) As the canoe progresses, its bearing in relation to the reference island changes. After traveling a certain distance on the ocean, the viewing perspective has changed, and the reference island now lies under a second star. The canoe has progressed to the end of the first *etak*.

As the canoe continues further, the perspective shifts again, and a third star comes to lie over the reference island. This is the end of the second *etak*.

The canoe is now traversing the third and final segment. As the destination grows nearer, a fourth star, the one that marks the reference island when seen from the destination island, replaces the third star. As the canoe reaches the beach, the fourth star will be moving into position, directly over the reference island.

It cannot be stressed enough that *etak* is not providing the navigator with a way to set or maintain a course. *Etak* is not "data." It is more like an operating system, running in the background of the CPU in a navigator's brain. Thomas Gladwin, the first outsider to write an authoritative account of *etak*, described it as a processor "into which the navigator's knowledge of rate, time, geography, and astronomy can be integrated to provide a conveniently expressed and comprehended statement of distance traveled," providing "the solution to an essential navigational question, 'How far away is our destination?' "

Etak is only as good as the raw data that lies behind it. "How does the navigator know where the reference island lies?" Gladwin wrote. "A study of *etak* will not give us the answer. Rather, we must look back to his instruction in the star courses." If the Polynesian navigator errs in his estimates of how far the canoe has traveled, *etak* will not save him and his crew. "The system is workable," Lewis concurred, "only because of the vast number of star courses and other items of information stored in the navigator's memory."

A navigator marking a position using *etak* is seeing the world very differently from someone who is not. To begin with, the canoe isn't moving. It is stationary on the ocean, watching the *islands* move. The island where the voyage began recedes, the destination moves closer, and the reference island seems to float between stable star points. "Everything passes by the little canoe," Lewis explained, "everything, except the stars by night and the sun by day."

It is a frustrating concept for us, and difficult for users to convey its workings to outsiders. It requires a worldview inconsistent with wayfinding as we understand it. Gladwin compared *etak* to looking at the world through the window of a moving train. Houses near the tracks seem to fly by, while mountains in the distance keep pace with the train over long distances. In *etak*, the canoe is the train and the stars the mountains. The stars are fixed in the sky. The islands, like the houses, are in motion.

When pressed, Carolinian navigators would always admit that *yeah, we get it—the islands aren't* really *moving*. But they would remain puzzled as to why anyone would consider this world-view problematic. For navigation purposes, they would explain, an island doesn't have one "location"—its position is relative to yours. "Strictly speaking it is not proper to speak, as I did, of the number of miles the navigator has traveled," Gladwin wrote of his difficulty fully comprehending *etak*. "In our speech we find it natural to estimate (or measure) distance in arbitrary units." For the *etak* user, "the estimate is relative. It is akin to a person walking across a familiar field in the dark. He is not likely to count his paces even if he knows their exact length. Instead he estimates intuitively that he is one-third or perhaps halfway across by knowing subjectively how long and how fast he has been walking."

Perhaps the European sea captain's image of the Pacific, as described by von Humboldt—islands with "no fixed position, [floating] from place to place over the charts"—was an accurate assessment. Von Humboldt was referencing the lack of accurate astronomical observations, but really, what does it even mean to place an island on a "chart," to see the spherical world through the distorted perspective of a standard map? "It is easy for us to forget, because of our familiarity, how much of an abstraction a chart really is," Lewis wrote. "The proverbial man from Mars would scan the ocean in vain if he expected to see marked there the same figures denoting fathoms and lines indicating shoals that a chart so prominently displays."

We all traffic in illusions when we try to define what it means to be somewhere, and not somewhere else. Societies rally around these shared conceptions of location. The home-center user thinks, *I am now 100 paces away from home, in the direction of the large tree.* The local-reference *etak* user orients himself by what segment of the journey he is on—it helps him know when to execute the steps in the course he has set. Even in the self-centered world, we maintain vestiges of both. In a strange city, with no map or GPS, we remember how many blocks we have ventured from the hotel; we locate ourselves in relation to prominent landmarks. But mostly, in the age of GPS, we don't require the environment to locate ourselves.

Imagine trying to convince a literal-minded Martian anthropologist—or the Carolinian navigator—why the first mapped image of the world many children study is the Mercator projection that is still a staple of grade-school classrooms, with its grotesque geographic distortions. Or why the X on a shopping mall's map, labeled "you are here," does not imply a belief that *you* are actually, you know, *there*, engraved on that map, anchored for eternity.

Lewis suspected that Tevake used a reference system similar to *etak*, one that allowed him to conceptualize where he was in relation to islands in the region. "His ability to point out the direction of invisible islands whenever he wished is presumptive evidence that he was thinking in terms of some form of home-reference system," Lewis concluded. But how did Tevake's system work? The concept had never quite broken through the language barrier—and the wayfinding barrier—so Lewis had never fully grasped it. Faced with Tevake's probable death, Lewis felt he'd lost not only a dear friend, but also a historical link. He described Tevake as "almost alone, one of the last in a line of navigators that stretched back 5,000 years to the Austronesian dispersal." "It is a matter for very real regret that our rather limited ability to communicate prevented me from questioning Tevake in any depth about this complicated subject, which seems to have been entirely neglected by earlier European investigators in Polynesia," Lewis wrote.

Whatever language barrier separated Cook from Tupaia and Lewis from Tevake was compounded by a cognitive barrier, an inability to reconcile two worldviews that help humans define location. Navigators using *etak* "share and take for granted all the cognitive antecedents of saying that an island 'moves,' " Gladwin wrote. "They find no need and therefore have had no practice in explaining to someone like myself who starts out thinking of a voyage as a process in which everything is fixed except the voyager."

If variations of this kind of local reference were once common throughout Polynesia, it would help explain how voyagers had such mastery of the sea. They were aided not only by ingenious navigation methods, but by a perfectly complementary mode of wayfinding that defined where they were, adding to that knowledge over the centuries, as the migration proceeded. And maybe, *just maybe*, this is what Cook had also missed when he showed little interest in Tupaia's map. It pained Lewis to think he had unwittingly repeated Cook's mistake.

Many who have pondered the mysteries of Polynesian migration, Lewis included, have expressed a disbelief, verging on outrage, that Cook expressed so little intellectual curiosity regarding Tupaia's map. If Tupaia was indeed trained to use a local-reference system, it would represent a worldview not amenable to expression in Cartesian space.

This cognitive barrier produces contrasting notions of the importance of timing for a navigator. Dead reckoning on the ocean would seem to require knowledge of both the average speed since the last known point and also how much time has passed. That temporal knowledge is absolutely essential if we want an accurate position fix. A skilled navigator using a modified home-reference system like *etak*—so adept at judging speed and thinking in terms of reference islands—has a more relaxed relationship with time. Lewis discovered that he was better than Tevake at estimating the exact time of day. But Tevake was so skilled at calculating the distance the canoe had traveled that he could estimate arrival time with astounding accuracy.

As it happened, Captain Cook was at the forefront of the effort to solve the timing problem. The eventual European conquest of the Pacific islands, accomplished in no small part by the refining of self-centered navigation, would decimate local populations. Among the losses would be the local-reference systems that abetted the original colonization. Already engulfed in that history, maybe Cook was incapable of really seeing Tupaia. And so their ships passed, moving apart like floating islands, Tupaia taking with him his perfect knowledge of the Pacific, and Cook looking away, in hot pursuit of the key to self-centering, a quest that would culminate in GPS.

CHAPTER TWO

The When and the Where

When James Cook visited Tahiti in 1769, the summer he met Tupaia, the captain was moonlighting as a stargazer. Every dozen decades or so, the planet Venus crosses the face of the sun—usually twice within a period of a few years—a phenomenon called the transit of Venus. In the late seventeenth century, the astronomer Edmund Halley had argued that close observation of the transit could help refine calculations regarding the distance between Earth, moon, sun, and other planets. These calculations would be more accurate, he noted, if observations were taken from around the world. Several worldwide expeditions had been dispatched for the 1761 transit, but the next, in 1769, involved a much larger global effort. Cook's would be one of about 120 simultaneous observations, organized by several countries, with the results compiled and analyzed by astronomers in Paris. Cook, ever the striver, knew that participating in the effort would boost his profile among British sea captains. His crew spent over a month preparing for the transit, scouting locations on Tahiti and even building a makeshift observatory.

Cook was interested in the transit for reasons beyond the gathering of scientific knowledge. European shipping was hobbled by an ongoing and seemingly insurmountable lack of positional awareness on the high seas. Latitude, a ship's north–south position, was a fairly simple proposition, calculable by observing astronomical phenomena such as the heights and angle of stars, using a sextant. Longitude, one's east–west position, was trickier, largely because it doesn't exist in an objective sense. It is more like a conceptual imposition engraved on the planet's face. The planet really does have an equator and two poles for us to define north and south. Longitude is defined in relation to a prime meridian, a line stretching between the poles. This meridian can be anywhere. Throughout history it has been placed at different spots, until the world reached a consensus that zero degrees is defined by a line running through the Royal Observatory, Greenwich, in London.

Simple astronomical observations won't yield longitude at sea. One needs to know what time it is at that spot on the ocean, and also what time it is at some spot where the longitude is already known. With that knowledge, the math is simple: the Earth rotates 360 degrees in twenty-four hours, so each difference of an hour represents a movement of 15 degrees from the reference point. A navigator could check the local time each day by noting when the sun was highest in the sky and calling that noon. The difficulty was maintaining the reference time. No clock existed that could withstand the humidity, salt air, changes in barometric pressure, and turbulence of a ship at sea.

Because of this lack of longitudinal awareness, navigators, no matter how detailed their charts, could never know with great certainty where exactly on the planet their ships were at any moment. To detect east–west position, they relied on dead reckoning techniques. They might compute speed by tossing a piece of wood overboard and timing how long it took to travel from one end of the ship to the other, and combine this data with a compass reading and other data to get a very crude calculation. Captains might increase the certainty of their bearings by hewing close to latitudinal parallels, the method Columbus used during his voyage to North America.

By the early eighteenth century, longitude miscalculations were responsible for several deadly shipwrecks. The problem was not merely safety, but also the perceived economic losses caused by the inability of ships to go anywhere and everywhere with confidence. The widespread skepticism that longitude was conquerable was reflected in the common colloquialism "discovering the longitude," which meant attempting the impossible. The longitude problem presented itself as a classic Enlightenment conundrum. To search for "the longitude"—and discover it—was to pursue a kind of perfect knowledge.

Conceptually, there were two potential solutions: either a clock robust enough for sea travel, or some kind of astronomical method of apprehending time. The overwhelming consensus was on the latter. The leading contender was designed by Nigel Maskelyne, England's Astronomer Royal. Using this method, called lunar distance, a navigator would determine Greenwich Mean Time by making celestial observations and consulting an almanac. The findings of the global transit of Venus observations was seen as a way to fine-tune lunar distance. But it was ultimately the horological solution that prevailed. John Harrison, a self-taught clockmaker, developed an ocean-hardened chronometer he called H4. The first reproduction of H4 was carried on Cook's second Pacific voyage, when his encounter with Tupaia was a distant memory. Having tested it against the lunar distance method for calculating longitude, Cook offered rave reviews of the chronometer when he returned to England in July 1775, calling it "our trusty friend" and "our never failing guide." The chronometer soon became standard. The longitude problem was solved.

The significance of the chronometer cannot be overstated. Its effect on the world rivals that of any other invention, including the printing press and the microchip. Dava Sobel, in her definitive history of Harrison's creation, *Longitude*, notes (without endorsing) the theory that the chronometer "facilitated England's mastery over the oceans and thereby led to the creation of the British

Empire, for it was by dint of the chronometer . . . that Britannia ruled the waves." Through the historical lens of Cook's voyages, the impact of the chronometer is conspicuous. Cook's first Pacific voyage opened the region to European exploration, even if just by affirming that the ocean existed as something more than a shipping lane; but his second voyage held the door open for others to rush in. The chronometer allowed Cook to make accurate shipping charts of the Pacific, which helped set in motion the processes of contact and assimilation that would roil the Pacific islands and enable the subjugation of its inhabitants.

The chronometer was the quintessence of self-centering. Not until the mid-twentieth century, when technologies developed during World War II evolved into widely used navigational systems, was the chronometer's importance eclipsed. Navigational systems such as radar, LORAN, and Omega all relied on some variation of timing the movement of radio signals to determine position. The question of *where* you are would always be conjoined with the question of *when.*

The transit of Venus may not have ultimately led to "discovering the longitude," but the mass effort to observe it in 1769, which involved nations dispatching observers to all corners of the globe, was the largest collaborative scientific effort humanity had ever attempted. It signaled a growing awareness of the power of simultaneity when trying to understand the many complex natural systems that drive the planet. In 1800, Chevalier de Lamarck, a French scientist, launched a five-year plan to solicit and analyze weather reports from around the world, to better understand storm formation and other meteorological phenomena. Nearly thirty years later, Alexander von Humboldt expanded on Lamarck's idea. To study the Earth's magnetism, he organized observatories across Europe and Asia, and later persuaded British and American scientists to join the effort, which led to the creation of the first international organization dedicated to compiling and sharing data relating to the earth sciences.

Scientists began to realize that studying Earth's polar regions yielded invaluable information about the planet's geomagnetism, as well as the complex processes that fuel storm formation and other meteorological phenomena. During the first International Polar Year project, which ran from 1882 through the fall of 1883, a dozen nations led expeditions to the Arctic and Antarctica. By 1932, when twenty-five nations participated in the second International Polar Year, the project had outgrown its name—only half of the 110 observation stations were located at the poles—but the goal remained the same: to understand the planet as a whole by gathering and analyzing simultaneous observations.

The next logical step was to take an even broader view of Earth. In 1950, Sydney Chapman, a British geophysicist, was invited to California to discuss atmospheric research with military officials. On the way, he visited the Applied Physics Laboratory at Johns Hopkins University, which oversaw several military-related projects. That night, at the home of the pioneering scientist James Van Allen, Chapman and Lloyd Berkner, the Pentagon's coordinator of scientific research, discussed a possible third iteration of the International Polar Year. To signal the project's expanded scope, it would be rebranded the International Geophysical Year, and last from July 1957 through the end of 1958, a period forecast to have large solar activity. Word quickly spread through the scientific community. John Simpson, a physics professor on the planning committee, would soon call it "the largest organized intellectual enterprise ever undertaken by man."

It was generally agreed that one of the major focuses of the International Geophysical Year would be the successful development and deployment of the world's first international satellite. The science of rocketry had developed quickly, but a rocket was of limited use to scientists, because it was only in the air briefly. If you could launch a satellite—and, just as important, track its position—it could be like an all-seeing eye, a reference point for transglobal observations. Perhaps in a nod to the era's burgeoning hi-fi craze, some scientists even referred to the satellite as a "long-playing rocket." Although this was a global project, there was a tacit understanding that the Americans would build the satellite. President Dwight Eisenhower vowed that the U.S. would have one in orbit by the end of the Geophysical Year.

The job was given to the U.S. Navy's Naval Research Laboratory, which dubbed the effort Project Vanguard. The work on Vanguard soon became the centerpiece of America's contribution to the International Geophysical Year. For the general public, Vanguard *was* the International Geophysical Year. The scientists basked in the glory. "Contemplate the satellite," mused Hugh Odishaw, who headed the U.S. Geophysical Year planning committee, "and you inevitably think about it in terms of yourself . . . of your destiny and transience of life." James Van Allen was awed "that we puny people can even contemplate hurling our own moon into the sky."

The head of the United States IGY Committee, Joseph Kaplan, declared the satellite would be "the greatest boon to astronomy since Galileo's telescope." Homer Newell, Vanguard's science coordinator, aligned himself with another historical figure: "This is our first step off the earth, and its possible significance is so staggering that I try to calm myself now and then by thinking of Columbus. For all he knew at the time he set out, he'd find nothing but man-eating sea serpents before his ships toppled over the edge of the world."

The Geophysical Year was a noble attempt to transcend Cold War politics. The project's rules stipulated that all work be public and transparent. The Soviets were slow to join the international project and remained coy about their plans, refusing to discuss whether their country had a satellite project similar to Vanguard. John Hagen, the head of Vanguard, insisted his team was not competing with anyone. At a press conference in November

1954, Eisenhower's press secretary was asked if he was concerned that the Soviets might win the satellite race. According to one report, he "snorted" his response: "I wouldn't care if they did."

If the politicians were bluffing, the scientists seemed sincere. On October 4, 1957, Berkner was at a Geophysical Year reception at the Russian embassy in Washington when he heard from a reporter that the Soviets had launched Sputnik. "I wish to congratulate our Soviet colleagues on that achievement," he said. Kaplan called the launch of Sputnik "really fantastic." The Vanguard scientists echoed this sentiment. They were impressed that the Soviets had launched a heavier satellite than their own, and into a more complicated orbit than the NRL team planned.

While the public was focused on the romantic idea of putting a satellite in orbit, some Vanguard scientists were tackling a more prosaic problem. A satellite in orbit had little practical research value without a way to follow and track its motion. This was no easy task. One Vanguard scientist compared it to following, from the ground, a golf ball ejected by a jet plane flying at 60,000 feet. The original Vanguard proposal contained plans for a system called Minitrack. An IBM computer at Cape Canaveral, Florida, the site of the satellite launch, would track its initial path, transmitting telemetry data to another computer in Washington, DC. That computer would calculate the likelihood of the satellite achieving orbit, and make predictions about orbital parameters. That data would zip by teletype to a network of fourteen radio tracking stations spread along the planet's 75th meridian, from Maryland to Chile. Whenever the satellite passed through the sky over a station, a sprawling network of ground antennas would measure the signal's angle. That information would be sent back to Washington, converted into punch cards, and programmed into the computer. Between seven to nine hours after liftoff, the computer would have enough data to compute the satellite's exact orbit and velocity.

Minitrack had another component. Based on the data, scientists at the Smithsonian Astrophysical Observatory in Cambridge, Massachusetts, would calculate where the satellite would be most visible, and when. Scattered around the globe, in Florida, Mexico, Iran, Japan, and eight other locations, observation stations were established, each equipped with a camera that could locate an object up to 500 miles away, linked to a clock accurate to within a millisecond. The observatory would relay the necessary information for stations within view of the satellite to calibrate the cameras. Project Vanguard had also initiated a program called Project Moonbeam that mobilized ham radio operators to track the satellite using a greatly simplified version of Minitrack designed by Easton, and the Smithsonian was putting together Project Moonwatch, providing instructions to help amateur stargazers search the sky for the satellite and note the time they saw it.

By the fall of 1957, Minitrack was nearly completed, ready to track any satellite the Geophysical Year would produce. But the Soviets had done something that baffled the scientists. Sputnik's signal was broadcast on a frequency different from the agreed standard for International Geophysical Year satellites. As news of the Soviet satellite spread, Vanguard's Minitrack team rushed to Building 72 at NRL headquarters, creating a command center and scrambling to gather information about the orbit which they relayed to personnel at the tracking stations, along with suggestions on how to adapt antennas to receive Sputnik's signal. At the stations, crews frantically soldered new parts and scaled the antennas with two-by-fours and wire cord to attach makeshift replacements. NRL also put out a call to ham radio operators worldwide to help get the signals to the National Academy of Sciences, where Berkner and others had set up an emergency room.

Meanwhile, the Smithsonian Observatory at Kittredge Hall was lit so brightly that emergency vehicles responded to reports of a fire. The cameras at the observation stations were not yet operational, so the SAO staff quickly put in a teletype and worked the phones. The observations from the worldwide Minitrack system were tentative, but good enough to allow the observatory to advise which teams could locate the satellite. The first observations that could be confirmed came in from Australia the next day.

Ultimately, only five of the stations along the 75th meridian caught the satellite. But under the circumstances, Minitrack performed well. It also planted a conceptual seed in the minds of the hundreds of amateurs who helped track the satellite. They, along with much of the public, evinced an attitude toward Sputnik similar to the scientists. It would take a few days before the real anti-Sputnik hysteria began.

Over the next few weeks, Sputnik came to represent an American failure. The space age had just begun, and already the country had fallen perilously behind. The Vanguard team's job now assumed a grimmer significance—though not necessarily for the scientists themselves, who maintained a pan-nationalist respect for their Soviet peers. A few months earlier, late in the summer of 1957, the Vanguard team had made plans to place a satellite on top of a rocket they called TV-3, which was scheduled for launch in December. TV-3 was one of a series of "test vehicles" for rocket design; the satellite was an afterthought. Maybe it would work, maybe not, but who cared? If it failed, they'd just keep trucking along through the International Geophysical Year, confident they'd eventually get it right.

By the December 6 launch date, however, anxiety about Sputnik had metastasized into a repository for America's wounded pride. The countdown began just after 5 p.m. on Thursday, December 5, 1957, and continued through the night. The slim emissary from America to the cosmos stood seven stories tall, but measured just 45 inches in diameter at its widest point. The TV-3 rocket balanced its payload atop this pencil-like shaft: a four-pound satellite the size of a grapefruit.

At 8:45 the next morning, the big red ball that signified an imminent launch went up over Cape Canaveral. At 10:30 a.m., an hour before liftoff, workers removed the red-and-white gantry crane that stood next to the rocket. With forty-five minutes to go, photographers skulked around the blockhouse, capturing the action as affirmative signals arrived from tracking stations around the world. J. Paul Walsh, Vanguard's deputy director, opened a phone line to his boss, John Hagen, a thoughtful, pipe-smoking man, who sat with four other Vanguard officials in a room at the Naval Research Laboratory. With a half hour left on the clock, loud blasts from the bull fiddle alerted everyone to clear the launch area. At the nineteen-minute mark, the room darkened and a "No Smoking" sign flickered on. The nervous engineers in the blockhouse stubbed out their cigarettes. A white trail of liquid oxygen shot out from between the first and second stages of the rocket. Five minutes to go: the countdown became audible on the public address system. (The engineer whose assistant did the counting down detected a quiver in the man's voice.) Walsh narrated the countdown over the phone: "Five . . . four . . . three . . . two . . . one . . . ignition. It's left the pad."

The sparks shooting out from the rocket's base became white flames as TV-3 began to rise. Three miles from the launch site, crowds along Cocoa Beach who'd gathered to watch the rocket were dazzled by the orange flames that cut through the late morning glare, but the veteran rocket-launchers knew something was wrong. TV-3 had toppled. The sound of ruptured fuel tanks carried for miles. The fireball was getting bigger. Someone in the control room yelled, "Oh God no!" One Vanguard engineer would later say the view out the window "looked as if the gates of Hell had opened up."

Word spread through the crowd. Workers at the luxury hotels along State Road A1A, places with proud names like Starlite and Vanguard, "stopped smiling," according to one report. "It was as if the region's pride had been deflated by the disaster." The deflation wasn't just regional—and it wasn't just symbolic.

The American economy—at least the part connected with Project Vanguard—began to contract. Within an hour, so many sell orders poured into the New York Stock Exchange from stockholders of the Martin Company, the major Project Vanguard contractor, that NYSE officials temporarily suspended trading of the company's stock. Martin wasn't alone. Investors ran from just about every publicly traded company involved with aircraft or missile materials that day.

On the phone with Walsh, Hagen was calm. Just make sure nobody's been hurt, he told Walsh, and make sure the press knows this. At a press conference convened a few hours later at a nearby Air Force base, Walsh's attempts at positive spin, emphasizing that the rocket's telemetry had functioned well ("It wasn't a long flight, but it was flying") did not go over well. With the wreckage of TV-3 still smoldering, the nation's editorial writers fanned the flames, issuing scathing missives that arrived the next day. (Adding insult to who-cares-that-there-were-no-injuries, this day was also the sixteenth anniversary of the bombing of Pearl Harbor.) Many assailed the atmosphere of hype that had dogged the launch since the day Sputnik went up. If we had done it low-key like the Russians, the logic went, we could have failed in profit without becoming a geopolitical laughingstock. "Pride goeth before a fall," fumed the *Hartford Courant*, "and the United States has just participated in one of the major pratfalls in history. Americans will have to grit their teeth in the days ahead and bear it." The *Richmond Times* called TV-3 "the most widely advertised flop of the decade."

Some of the commentary displayed a hostility to the military that would be unthinkable today. Under the headline "Goofnik Blows Up," the New York *Herald Tribune* requested that from now on, Washington "damn well keep quiet until they have a grapefruit or at least something orbiting around up there, and until they do, just shut up." Such was the paper's anger that it approvingly noted that the Soviets "have a habit of liquidating their bureaucrats who fail. Some heads ought to be rolling over here, too." Another New York paper, the *World-Telegram and Sun*, weighed in to say that the *real* tragedy was not "that the coffin varnish and tabasco sauce or whatever it is that fueled our rocket exploded today." It was "the hopeless hand-wringing" of bureaucrats. "Even more than the beep-beep of our own Sputnik, the sound we most want to hear is the thud-thud of heads being knocked together in Washington."

The Vanguard team, so recently revered, were the goats of rocket science. Lyndon Johnson, chair of the Senate Preparedness Committee, called the situation "most humiliating," and vowed a "full, complete, and exhaustive inquiry" into Vanguard. Over at the United Nations, the Russian delegation waggishly inquired if the U.S. was interested in a Soviet program that offered technical assistance to backward nations. Homer Newell, who had so recently invoked Columbus, now chastised his fellow Americans who "think of science in terms of applications—things like fabrics and tail fins—not a patient search for knowledge for its own sake."

The funny thing was, the grapefruit was mostly intact. It had been thrown clear of the wreckage, and was still broadcasting its beacon from the pavement. Roger Easton, a co-author of the Vanguard proposal who enjoyed tinkering with the TV satellites at his family's dining room table, gathered it up and put it in a little box in his office. Easton, thirty-six years old, was neither defensive nor contrite—perhaps because Minitrack had gone well, right under the nose of the public. He claimed that he and his colleagues found it ludicrous that the toppling of TV-3 was construed as this catastrophic. "A new rocket," he pointed out, "is just as apt to blow up as to go up."

The world learned about Sputnik on a Friday. When William Guier and George Weiffenbach, two young engineers at Johns

Hopkins' Applied Physics Laboratory, arrived at work the following Monday, they were surprised to discover that nobody there had tried to receive Sputnik's signal. This was an odd dereliction—or at least a demonstration of the sanctity of the weekend for rocket scientists—given APL's tight ties with the military as a major defense contractor. Also, it was APL alum James Van Allen who had provided the initial intellectual catalyst for the International Geophysical Year, the endeavor whose thunder Sputnik had stolen.

The young engineers figured that, if nobody else was on it, they might as well try to do it themselves. As it happened, they had the perfect skill sets. Guier had a background in computer science. He had recently conducted hydrogen bomb simulations using a supercomputer for the Atomic Energy Commission. Weiffenbach was writing his PhD dissertation on microwave spectroscopy, a technology that the two discovered could be very useful in not only receiving Sputnik's signal, but also recording it.

Guier and Weiffenbach found they were able to tune a receiver to Sputnik's frequency, and by late afternoon, they were picking up steady blips. The next step was to record the signal. Using a high-fidelity tape recorder Guier had recently purchased, they recorded the sound the signal made as Sputnik traveled from horizon to horizon. The beacon's rhythm was constant, but the sound varied. As the satellite came over the horizon, the note went higher, peaking as it passed overhead, and then grew steadily lower as it receded. This was the Doppler effect, the frequency change a stationary observer experiences when perceiving waves emitted by a moving object.

Because the Doppler effect is uniform, stable, and predictable, it can be used for measurement. A microwave signal aimed at the moving object will bounce back. Analyzing the frequency change in the reflected signal yields information about the object's motion, such as its velocity and heading. Guier and Weiffenbach discussed how they might measure Sputnik's signal by adapting a method recently developed by APS for tracking guided missiles. In the days that followed, Weiffenbach worked on digitizing the Doppler signals, a painstaking process in the earliest days of digital computing. Guier crunched the numbers to compute Sputnik's nearest approach, the point at which it was directly overhead, when the tone would be at its steadiest. By the end of October, when Sputnik's batteries were depleted and its blip had gone silent, they were making accurate predictions about Sputnik's orbit.

In March 1958, Frank McClure called Guier and Weiffenbach into his office and asked them to shut the door. The first person to head APL's Research Center, a position he had held since 1949, McClure was barely in his forties. He was renowned for his ability to conjure practical applications from the breakthroughs of his charges. Now he had an idea he wanted to bounce off the young scientists. They had proven that a stationary observer, using Doppler data, could measure the behavior of a moving object in space. That was an achievement in itself. But couldn't the inverse also be true? If you knew exactly where that moving object was, at any given moment, wouldn't that tell you the exact location of the observer on the ground? Guier and Weiffenbach, enthralled by the gee-whiz experimentation of their project, had not noticed its larger significance. They had, almost accidentally, invented the world's first satellite navigation system.

McClure knew that the Navy was looking for an effective way to position Polaris nuclear submarines. At the time, the most common method was to use LORAN (long-range radio navigation), a positioning technology that employed land-based transmitters. The biggest problem with LORAN was that, as a terrestrial system, its transmitters were vulnerable to signal-jamming and physical attack. The Doppler-based system McClure envisioned would be more secure, since the transmitters were in space. It would also be a *passive* system—obtaining a position fix would only require a user to receive a signal, not send one.

Following the meeting with McClure, this Doppler-enabled satellite navigation system came together remarkably quickly. The next day, McClure wrote a memo to his boss, APL director Ralph Gibson, explaining that while speaking with Guier and Wiffenbach, "it occurred to me that their work provided a basis for a relatively simple and perhaps quite accurate navigation system." Gibson then told the Navy's Chief of the Bureau of Ordnance that he was confident APL could design a navigation system that was accurate to within a half mile. APL threw together a fifty-page outline of the Navy Navigation Satellite System. In less than three weeks, Guier and Weiffenbach's pet project had gone from geeky science experiment to the skeletal frame of the world's first satellite navigation system.

The program, which soon came to be known as Transit, began with a budget of just $1 million. Tracking stations in the continental U.S. and Hawai'i picked up the Doppler data from Transit satellites, and transmitted them over land lines to a computing center in Southern California, which determined the orbital parameters of each satellite. All of this information was sent to "injection centers" that transmitted it back to the satellites, which in turn broadcast it to anyone with a Transit receiver. Every receiver generated a reference signal that was compared to the signal received from the satellite, producing a "Doppler count" that the receiver used to compute its location.

The Transit system was fully operational by 1964. Three years later, Vice President Hubert Humphrey declared that Transit was available for civilian use. The first nonmilitary users were oceanographers on research vessels—and soon there were 80,000 Transit navigation units in use, for everything from commercial shipping navigation to surveying to helping establish the boundaries of oil and mineral deposits.

When Transit was finally decommissioned in 1996, one of its former program managers memorialized it as "the largest step in navigation since the development of the shipboard chronometer." It turned out to be a prelude to an even larger step: the development of a new chronometer. The eighteenth-century sailor had struggled to perceive time on the ocean. Space was the next frontier.

When Transit launched, Roger Easton was heading up the Naval Research Laboratory's Space Applications Branch. He was pondering a conceptual inversion similar to what McClure saw in his young engineers' invention. Minitrack was a success, but its utility was limited. It could only track satellites, like Sputnik, that emitted a clear signal at precise intervals. A Soviet spy satellite would not be so obliging, and would pass undetected.

NRL's solution was to install a network of massive horizontal antennas, some stretching as far as a mile, at stations along the 33rd parallel, from Southern California to Georgia. These antennas beamed radio waves skyward to form a radio "fence," its east–west inclination deemed ideal for detecting foreign spy satellites launched into north–south "polar" orbits. Any airborne object as small as a basketball, at an altitude of up to 15,000 nautical miles, would pass through the fence. Analysts could infer information about an object's position and orbit, but not much about its speed and direction. For that, they needed a second radio fence, which they built more than 100 miles south of the first. A satellite would now penetrate both fences, yielding a much better picture of its behavior. The system was given the unwieldly acronymic name NAVSPASUR (Navy Space Surveillance System).

But this arrangement presented another problem. The two-fence system's effectiveness in measuring velocity depended on the existence of two synchronized clocks, one at each station. The clocks would require periodic resyncing, which meant someone had transport one clock to the other station, physically connect the two clocks, and then make the hundred-mile return trip.

It was in September 1964, at the NAVSPASUR station in south Texas, that Easton began seriously thinking about an idea he had been kicking around since the previous spring. The main problem with the Transit program, he had noted in a technical memorandum in June, was that it was "not general enough," focusing on Doppler navigation when there were so many other avenues to explore. In particular, he wanted the Navy to investigate the concept of "range measuring" as the basis for satellite navigation. "The idea is not new," he wrote, "but now the art is developing to the point where it appears to be feasible."

What if an orbiting satellite, traveling with a highly accurate clock, and with a clear sightline to both clocks on the ground, could synchronize the clocks by sending them a time signal? And just as McClure had when he thought of measuring the Doppler shift, Easton realized that if this time system were possible, it had the workings of a passive navigation system based on time signals. What Easton was conjuring, on that dusty day in the bosom of the Rio Grande Valley, was a skeletal version of GPS. Easton called the idea Timation, a portmanteau for "time navigation."

The general principle is called passive ranging. Imagine you and a friend who lives far away each possess highly accurate synchronized clocks. In addition to your clock, you have a live video feed showing the face of your friend's clock. You notice, from looking at the feed, that your friend's clock is just slightly off from yours. What does this tell you? Perhaps one of your clocks is malfunctioning. But if you can rule out that error, and know with absolute confidence that both clocks are working perfectly, this discrepancy becomes *information*. The lag is caused by the time required for the image of your friend's clock, traveling at the speed of light, to reach you. The speed of light is constant and stable. Your clocks are constant and stable. The lag is directly related to the distance between you and your friend. You now have tools in place for a satellite-based passive positioning system.

Maintaining perfect time on a satellite posed an even greater challenge than on the sea—but an ongoing revolution in timekeeping offered a way forward. Until recently, the world had defined time as a function of the Earth's orbit around the sun (solar time), or the movement of celestial bodies with respect to the Earth (ephemeris time). The instability of both methods made ultraprecise timekeeping difficult. Atomic clocks offered the opportunity to change the time scale from something governed by behavior outside the clock, to something located within the clock itself. The oscillations of atoms are inherently stable, so a second can be defined as the time required for an atom to go through a certain number of cycles of vibrations.

Within an atomic clock, microwaves make electrons bounce between two states, and these changes govern the clock's timekeeping. The idea was discussed as early as the 1920s, but the earliest prototypes did not appear until twenty years later. The timing community settled on caesium as the atom of choice for atomic clocks, and by the mid-1950s some clockmakers were experimenting with designing caesium beam clocks that were both portable and commercially available.

Until the development of atomic clocks, the most accurate timepieces used oscillators built around quartz crystals. Highly piezoelectric, quartz generates an electrical charge on its surface when mechanically stressed. In the reverse process, employed for oscillators, quartz bends in response to a charge, vibrating like a tuning fork at a very high and stable frequency. Circuits tick off seconds based on the number of vibrations. In 1960, Harvard researchers developed a frequency standard called hydrogen maser. The process involved bombarding hydrogen atoms with microwaves that made them flip from one state to another, at a rate predictable enough to stabilize a quartz oscillator.

It was the hydrogen maser that ultimately convinced him of the practicality of a positioning system based on passive ranging. Atomic clocks were still too bulky to be useful. But a clock with a standard oscillator kept in lockstep by an atomic standard might be hardy enough for space travel.

In lieu of a satellite, Easton devised a ground-based experiment. He modified a receiver at the south Texas station so that it emitted a steady sequence of tones. An engineer named Matt Maloof put a similar receiver in a convertible he owned. Maloof got in his car and proceeded to drive down an unfinished stretch of Texas state highway. The system measured the signal as it flew out to Maloof's receiver and back, and calculated the distance from the station to Maloof. With the pedal down and the hot Texas wind in his hair, Maloof flew down the deserted road. Back at the station, the numbers crept upward. The system worked.

Easton staged something similar for the brass back in Washington. Maloof again volunteered his services. "We had invited a bunch of big-shots that had money, and said, 'Okay, here's a demonstration of passive ranging,'" recalls engineer Pete Wilhelm. "His convertible goes racing down the highway, and we were able to constantly measure the range between the receiver on the building and the car. The plot was very, very, beautiful. You could see that there was very little noise. We knew exactly where the car was." As they looked at the screen, they were amazed to discover they could see when the car changed lanes.

Easton encountered some resistance regarding the workability of a full-fledged navigation system based on passive ranging. "Some time was spent in explaining the principles of operation to various parts of the Navy," he wrote in a memo outlining progress on Timation in 1967. "It was finally resolved that perhaps we did know what we were talking about. The good result of this confrontation was that we never again had much trouble with arguments concerning the system. We knew all of the existing problems and most of the nonexisting ones."

One of the most delicate problems was political, not technical. Transit, a system built by another segment of the Navy, was already operational. The Naval Research Laboratory would essentially be competing with its own branch of the military. Transit was optimized for oceangoing vessels, which could remain in one place for the fifteen minutes or so required to get a position fix. It was of little use for pilots. "With the Doppler thing, the restriction on users was rather significant," says Pete Wilhelm, an engineer who worked on Timation. "You had to be standing still, because if you're moving, you can't separate your motion from the motion of the satellite. The faster you're moving, the more inaccurate your location is going to be."

Despite the improvements over Transit offered by Timation, Easton was given an initial budget of just $35,000, the maximum amount Timation could receive without Easton lobbying further up the chain of command. To really test the system, Timation needed multiple satellites. They could barely afford one. Wilhelm, who had never worked with rockets before, conceived of a way to use decommissioned Atlas ballistic missiles, which were gathering dust in California, to carry a satellite partway into space, providing enough of a boost for the satellite to continue to its orbital height.

In 1967, the world's timekeepers voted to adopt atomic time as the standard. The second severed its link with the Earth's rotation or any other astronomical phenomena, and would now be "the duration of 9 192 631 770 periods of the radiation corresponding to the transition between the two hyperfine levels of the ground state of the caesium-133 atom." That same year, Easton's groups launched its first satellite, Timation I, orbiting at 500 nautical miles, carrying a clock with a crystal oscillator accurate to one second every thousand years.

The navigation tests NRL conducted with boats, trucks, and planes were promising, but the ionosphere wreaked havoc with the satellite signal. Timation II went up two years later, with better equipment that could withstand the atmospheric problems. With this satellite, NRL finally tested Easton's theories of satellite time transfer, attempting to synchronize two of the world's most accurate clocks, those at the Naval Observatory in Washington and the Greenwich Observatory in London. The goal was to have the satellite's clock, which received its time signal from the Naval Observatory, transfer that time to Greenwich, so that all three clocks were in sync. For a week, every time the satellite came over the horizon, staff at both places gathered clock data, sending it to computers in Maryland and Virginia for processing. A General Electric time-sharing computer in Cleveland served as the communications link between the two observatories. The groups later concluded that the satellite transfer had synchronized both clocks to within a few ten-millionths of a second.

The navigation experiments were not quite as successful—the satellite's clock was again vexed by atmospheric conditions—but Timation II still provided position fixes to within 200 feet. Work was underway on Timation III, which was slated for a much higher orbit than the previous two—8,600 miles high, providing larger global coverage—this time with an onboard atomic clock. Responding to a directive from the Joint Chiefs of Staff, NRL began plans to expand the Timation concept to include a constellation of satellites. The Timation team had concluded that ranging to four satellites simultaneously would provide highly accurate positioning in three dimensions.

In the post-Sputnik years, the imagining and building of satellite navigation systems occurred on a few fronts. The same year NRL launched Timation I, the Soviets launched the first satellite of the Tsikada program, a Doppler-based system similar to Transit. The U.S. Air Force was also conducting research on passive ranging, under the aegis of a program called 621B. Like Timation, 621B could not test its theories with a full constellation of multiple satellites, but while Easton's team concentrated on the timekeeping facet of this kind of system, the Air Force group simulated the effects of such a system by using fake satellites ("pseudo-lites")—four stationary transmitters planted in the ground at the White Sands Missile Range in New Mexico, arrayed to mimic a satellite constellation and emitting a synchronized time-coded signal. An airplane equipped with a receiver to pick up the signal was navigated around the desert using the pseudo-lites. The results showed that the 621B concept could achieve an accuracy of five meters in three dimensions.

As satellite systems that utilized passive ranging, Timation and 621B both offered a superior alternative to Transit, and were similar enough that only one could continue to receive funding and support. Both showed promise, but neither program had fully progressed past the proof-of-concept stage. The Pentagon, reluctant to take sides, let the issue fester. Meanwhile, other ideas for satellite navigation systems were percolating throughout the armed forces—"literally hundreds," according to Brad Parkinson, the Air Force officer who assumed control of 621B in 1971—but it was clear that the Pentagon would ultimately have to choose between Timation or 621B, or perhaps just stick to improving the already up-and-running Transit.

Parkinson would soon witness the summary execution of 621B. From its ashes, he would raise a passive ranging satellite navigation system. But its purpose would be anything but passive. The man who breathed life into what became GPS wanted to build a new and improved way to bring death from above.

CHAPTER THREE

Global Reach, Global Power

The United States Air Force never really wanted GPS. The 621B program, the precursor to GPS, was underfunded. After it evolved into the GPS program in the early 1970s, the Air Force largely neglected it, to the point of disowning it and defunding it. A few times, it tried to kill its own creation, and GPS was kept alive by the Pentagon's largesse. This is difficult to understand, especially in light of the post-Sputnik race for space, the geopolitics of the Cold War, and the global saturation of GPS today. Who wouldn't want GPS?

One reason the Air Force was slow to embrace GPS is that space-based projects were never seen as a priority. "The Air Force is not a big user of space," says Scott Pace, who tracked early GPS use from within the U.S. Department of Commerce, and now directs the Space Policy Institute at George Washington University. "The Air Force gets to *build* for space, but the Marine Corps, Army, and Navy are much more reliant on actual space services than the Air Force itself is. The budget for space is in the Air Force, but in terms of the number of customers and users, they're all in the other services. So there's always been a tension."

For many, the utility of GPS was not apparent. The problem of navigation seemed largely solved. "You don't know how many times I heard, 'What does it do?' " says Gaylord Green, one of the small group of Air Force officers who designed GPS. " 'It tells you where you are.' 'I know where I am, why do I need a damn satellite to tell me where I am?' " Ron Beard, Roger Easton's number-two man on the Timation project, recalls a similar refrain from Navy colleagues who were not part of the Naval Research Laboratory: "We're the Navy, we know where we are."

"When I was at the Pentagon, every year it was my job to defend the GPS budget to the Air Force," says Ronald Yates, a retired Air Force general. "And every year, the operational command zeroed it. Every year the issue was the same—I got sick of hearing, 'We don't need another navigation system.' And they were right. They *didn't* need another navigation system. But my point was, it isn't a navigation system. Think of it as a *guidance* system." That is a subtle yet very important distinction. Until GPS became fully entrenched, its biggest supporters envisioned it not so much as a navigation or wayfinding tool; they saw it as a way to drop bombs and launch missiles.

"The Air Force was conflicted back in those days," says Jules McNeff, the chief officer responsible for defending the budget for the GPS program in the late 1980s. "GPS was this brand new space thing that hadn't developed a constituency. Every dollar spent on GPS was one dollar less spent on some piloted aircraft. And the Air Force is run by pilots." As Yates and McNeff make clear, perhaps the greatest hurdle facing GPS was the Air Force's corporate culture. The majority of the Air Force is on the operational side, those who either fly aircraft or facilitate the flying of aircraft. A much smaller minority fall under the space heading, which is viewed with suspicion by the operational side. "The bottom line," Green says, "is you got airplanes and you got satellites. And the Air Force is pilots who fly planes." Chuck Horner, the Air Force general who oversaw the air campaign during the Gulf War, is even more succinct: "They're paid to dream. We're paid to kill."

With Horner's blessing, GPS would play an important part

in the air campaign that began the Gulf War. GPS would prove even more instrumental in the success of the ground offensive that followed it. The Gulf War was GPS's debut on the world stage, when it became utterly impossible for the military to dismiss its utility. It was the culmination of a journey begun by yet another Air Force officer, who had become convinced that GPS offered some antidote to the excesses of Vietnam. And he occupied a special place in the Air Force's institutional divide. He was a dreamer who found a better way to kill.

It seemed to Brad Parkinson as though his entire academic and professional life was a path that led to GPS. As a cadet at the Naval Academy, he studied electrical engineering, researched navigational techniques, and learned about precision weapons delivery systems while doing summer cruise training on the battleship *Missouri*. Upon graduating, he entered the Air Force to concentrate on controls engineering. The Air Force offered to send Parkinson directly to do graduate work at MIT, but first—"to find out what the Air Force was all about," as he recalled—he spent two years learning aircraft electronics and serving in an operational squadron. At MIT, he studied with Charles Stark "Doc" Draper, one of the world's leading authorities on inertial navigation. After two years doing guidance and navigation analysis for the Air Force, he earned a doctoral degree in astronautics from Stanford. A short time later, he joined the faculty at the Air Force Academy. Soon after arriving, he was given a special detachment that placed him at the heart of a debate that had roiled the Air Force from its earliest days, and attained a renewed urgency with the escalation of the Vietnam War.

The question revolved around the ethics and efficacy of so-called strategic bombing, air campaigns that targeted nonmilitary targets such as factories, infrastructure, or densely populated areas—to disrupt the enemy's ability to fight or to weaken the resolve of its citizens. As early as the 1920s, the United States Army Air Corps preached a doctrine of high-altitude precision bombing aimed at industries that supplied the enemy, carried out in the daytime for maximum visibility. The problem was that daylight made the bombers more vulnerable to attack. The solution was the B-17A Flying Fortress, introduced in 1935, and integrated into what was by then called the U.S. Army Air Force. The aircraft was equipped with machine guns, heavy armor that obviated the need for fighter escorts during bombing runs, and a harbinger of the future of warfare: a bombsight with an electromechanical bomb release calculator, the first precision aiming system designed for aircraft.

The B-17A proved to be anything but a precision bombing panacea. The aircraft's excess armor and armament limited its range and bomb load capacity, and the introduction of radar drastically reduced whatever safety there had been in remaining at high altitude. B-17s suffered heavy losses in air campaigns over Germany, including a disastrous mission in 1943 that targeted factories in Schweinfurt and Regensburg, in which sixty B-17s were destroyed.

Unescorted high-altitude daylight bombing fared no better in the Pacific theater. B-29 bombers were flummoxed by bad weather and heavy winds that compromised their bombsights, with some raids on Japanese factories producing not one direct hit. Existing bombing systems were simply incapable of precision targeting. It wasn't uncommon for bombs to miss their target by more than a mile—and even that level of accuracy required hazardous daytime raids.

By 1945, despite significant resistance within the Army Air Force, the precision ideal was abandoned in favor of wide-area incendiary—and largely indiscriminate—attacks on Japanese cities, whose structures, largely comprised of highly flammable material, resulted in apocalyptic firestorms. Low-altitude nighttime bombing of urban areas, which began with a raid on Tokyo in March 1945, had killed more than a quarter million people by June, left nine million homeless, and obliterated two million buildings. Not that it mattered, but nearly 90 percent of the bombs had missed their intended impact points by at least 2,000 feet.

"What happened in World War II was a travesty," Parkinson says. "There was no precision weapon delivery. Bombs were delivered helter-skelter everywhere. They were as much an element of terror as an element of actually destroying things." The Air Force clung to this approach in Vietnam. "They were accustomed to the World War II tradition of carpet-bombing," he says.

By the time the Air Force entered into the brewing conflict in Southeast Asia, it had been an independent branch of the military for less than two decades. For the first several years of its existence, the Air Force had emphasized a continuation of World War II-era bombing strategies. A small but vocal contingent within the service argued that wars in the postcolonial era would increasingly emphasize guerrilla warfare, and that the Air Force would be more effective if it revisited the idea of targeted bombing. An Air Force officer named Ron Terry advocated for a new version of an old concept: the fixed-wing gunship.

The gunship concept extends back to the early days of aviation. Pilots delivering mail and supplies to remote regions such as the Amazon or the Australian outback learned that if they tied their package to a rope dangling from the side of the plane and flew the plane in a pylon turn—a continuous orbit on an imaginary axis extending from the plane to a single point on the ground—the package would hang in one place, making it easy for someone on the ground to retrieve. A fixed-wing gunship has armaments that fire from the side of the aircraft as it makes a pylon turn. Done right, a gunship allows for precise targeting from a fairly high altitude. The reality of engaging the Viet Cong's guerrilla tactics was that aircraft were spotting targets on a first pass and then losing them on the second. Fast-moving jet aircraft tasked with supporting ground troops were missing targets, and sometimes even dropping napalm on their own soldiers—and they had no all-weather or night capabilities.

After introducing one new gunship into the war—the AC-47, a modified version of Boeing's commercial DC-3 plane—the Air Force next introduced the AC-130 gunship, a transformation of the C-130 transport aircraft. The AC-130 was like something a rich playboy-turned-masked-superhero would design in his basement lair: side-looking and forward-looking radar, two 20-kilowatt xenon arc lamps that gave off infrared and ultraviolet light, Doppler radar for navigation, a semi-automatic flare dispenser, homing instruments, and a computerized firing system that linked the guns—eight of them, capable of firing thousands of high-explosive incendiary shells each minute—with sensors. A shooter could aim, fire, and hit a target without ever actually *seeing* it.

The first prototype AC-130 entered the war in February 1968, and was used mostly for interdiction efforts (disrupting the movement of troops and supplies) along the Ho Chi Minh Trail in Laos. In November, Lyndon Johnson announced the cessation of much of the bombing of North Vietnam, a move that shifted more resources to the interdiction campaign while simultaneously increasing the dangers for air patrols, as North Vietnam began moving more anti-aircraft guns to the trail. Although only four AC-130s were in combat, by April 1969 they were responsible for nearly half of the trucks destroyed or damaged in the area. President Nixon, responding to the Tet Offensive, announced his plan for "Vietnamization," the gradual withdrawal of U.S. ground troops. But the air war continued, especially over the Laotian panhandle, where interdiction efforts became even more heated. Nixon's secretary of the Air Force said he worried about the vulnerability of the AC-130, especially after one was shot down, killing two on board and destroying a quarter of the fleet. In July, Terry and some others who were involved in designing the prototype AC-130 met to discuss the aircraft's future, bringing in advisors from the Air Force Academy. One of them was Brad Parkinson.

In October 1969, just after Parkinson arrived at the school, William Westmoreland, the Army's Chief of Staff, predicted that within ten years wars would be waged using "the automated battlefield," the ability to "destroy anything we locate through instant communications and the almost instantaneous application of highly lethal firepower" using "hundreds of surveillance, target acquisition, night observation, and information processing systems."An embryonic version of this automation already existed in the form of Igloo White, a $3 billion Air Force program that involved gathering intel by dropping sensors in the jungle (so sensitive they could supposedly detect the smell of urine), transmitting the data via drone aircraft to a control center for processing, and then to manned aircraft, including the AC-130s. But for Parkinson, it was the AC-130—"without a doubt, the most precise weapon delivery system we had in that era"—that embodied this new era of warfare, because the aircraft represented an alternative to the Air Force's carpet-bombing ethic.

Parkinson had barely settled in at the Academy when he was asked to join the AC-130's development team. He spent several hectic months working to perfect the digital control system that calculated and directed the airplane's line of fire. After successful Stateside tests, this first modernized AC-130 was deployed to combat in Southeast Asia, with Parkinson a member of the combat crew. For four months, he flew on nearly every night mission, spending four to five hours operating the fire-control system to see how it performed in total darkness. After twenty-five missions and about 150 hours, he remarked that the experience—for which he received a Bronze Star—gave him a "keen appreciation" for "what it was like to get shot at" and "the value of precision weapons delivery."

The lessons stayed with him when he returned to the U.S.

and agreed to helm the fledgling 621B program. The appeal of a very precise passive ranging system, as a military tool, seemed obvious. Parkinson assembled a team of around twenty-five Air Force officers with engineering backgrounds, convening them for daily early-morning technical meetings at Los Angeles Air Force Base. In the spring of 1973, he found an unexpected ally. Malcolm Currie, the third most powerful person at the Department of Defense, was in the process of relocating his family to Southern California, often flying there for the weekend. He often dropped by the base in El Segundo on Fridays to receive briefings from Kenneth Schultz, the soon-to-retire commander of the Air Force's Space and Missile Systems Organization. One such Friday, Schultz, having run out of things to discuss with Currie, suggested that Currie meet the young colonel working on building a satellite navigation system.

Parkinson suddenly found himself face-to-face with the number three man at the Pentagon. Currie asked him what 621B was all about. Parkinson, quickly regaining his composure, delivered an impromptu discourse on the program: passive ranging, precision weaponry, how the project would come together, and the finer points of how the technology functioned. Currie, who had an advanced physics background, was captivated by the ideas Parkinson presented. After three hours of discussion, it was clear he understood the possibilities inherent in the cutting edge of navigation, positioning, weaponry, and perhaps even timing.

In August, Currie had Parkinson visit the Pentagon and present 621B to the high-level committee that would need to approve its further funding. Parkinson made an impassioned case for the project, but to no avail. Almost immediately after he concluded his remarks, the group, which had apparently reached its conclusion before hearing a word, voted to kill the project. Currie invited his crestfallen protégé back to his office, and helped him hatch a new plan.

Parkinson could propose the formation of a Joint Program

Office, overseen by the Department of Defense and managed by the Air Force, but including representatives from other branches. Currie felt that the program would be strengthened by having diverse stakeholders, although this would probably be perceived within the Air Force as a threat to its autonomy. "I recognized that there was pressure on me from the Air Force to do it the Air Force's way, so I came up with the idea that we would have a synthesis of all the ideas and some of our own," Parkinson recalled.

On Labor Day 1973, when the massive institution was all but deserted for the long weekend, Parkinson convened a group of about a dozen people at the Pentagon to hash out details of a new global satellite navigation system. In one of the few lit rooms in the otherwise dark Pentagon—Parkinson would later canonize this gathering as the "lonely halls" meeting—Parkinson and his handpicked team of personnel representing the Air Force and Aerospace Corporation, an Air Force-affiliated think tank, sketched the basic outline of what was to be called the Navstar Global Positioning System. (The first word, used sporadically for a few years, was eventually dropped altogether.)

One of the most controversial aspects of the system was the decision on how to structure the ranging signal sent by the satellites to GPS receivers. Spread-spectrum technology, of which the GPS signal is one form, has an unlikely provenance. In the early 1940s, the actress Hedy Lamarr, at the pinnacle of her Hollywood fame, collaborated with the composer George Antheil on a wartime idea they believed would prevent jamming of the signals sent to radio-controlled torpedoes. They proposed spreading the signal over several different frequencies—effectively increasing the signal's bandwidth—so that an enemy would have the difficult task of jamming them all to prevent the signal's informational content from getting through. (Lamarr's and Antheil's original control device for randomly switching frequencies, never put into practice, was a player-piano roll.)

Spread-spectrum systems generate the type of coded signal

that cryptographers call pseudorandom noise, because it appears to have no pattern, and therefore no capacity to communicate information. In truth, there is a pattern, with the garbage surrounding it serving as a layer of security. Anyone authorized to receive the signal's message has a code generator that follows the same rules. By comparing its own generated code with the code transmitted by the sender, the receiver can locate the message and ignore the rest.

This is what every GPS receiver does. The GPS satellites transmit their pseudorandom digital code; 67 milliseconds and 20,000 kilometers later it reaches us, allowing any GPS receiver in its path to extricate the message from the babble around it. The message describes which satellite sent the signal, the time the signal was sent, and other information about the current positioning of the entire constellation. By ranging to at least four satellites, the receiver performs its calculations to its fullest capacity.

That 20,000-kilometer journey is treacherous. When the signal reaches Earth, it is almost impossibly faint, virtually indistinguishable from the electronic crackle that blankets the planet. Rescuing it is roughly comparable to reading a book in London by the light of a 10-watt bulb in Rome. The GPS receiver misses some of the message component, but picks up enough to piece together the rest. In doing so, it filters out the noise, in essence making the message "louder." The GPS chip in your smartphone performs this task well enough to amplify the signal by a factor of one million.

The specific kind of spread-spectrum method used by GPS (as well as some mobile phone carriers) is called Code-Division Multiple Access. CDMA allows all the GPS satellites to transmit on the same frequency, each identified by a unique pseudorandom code. Magnavox, at the time a small Bay Area startup, created CDMA, primarily as an espionage tool. "It was a secret technology we'd been working on since the late fifties, early sixties," Len Jacobson, the director of business development at Magnavox during that period, recalls. "The idea was to be able to get airplanes that were being jammed into Berlin. We were building modems that used spread-spectrum, and it was all a secret. And then GPS came along, and we basically took the same signal and said, 'This will be the GPS signal.' Parkinson sort of gave it to the world."

Attempting to build GPS on the back of such a faint signal struck many outside the Joint Program Office as sheer brazenness. The Army expressed concern at how the signal would behave in heavy cover. "I remember walking into the meeting—we were about two years into the satellite development," says Gaylord Green, who supervised that segment of GPS. "Those guys had a 'tree model' that showed as you went through a forest, how the losses went up until it was impossible to track. They said, 'We need more power out of the satellite!' I said, 'Well, you're going to have to tell the Army to find a clearing.'" Green laughs at the memory. "That didn't go over very well," he says. "And I always felt bad about that—particularly when I'd drive under a tree and lose GPS!"

Sometimes it seemed like people went out of their way to remark on what a disaster GPS would surely be. One day at the Pentagon, an Air Force general walked right up to Parkinson to inform him that Parkinson's "otherwise brilliant career was going down the toilet" if he insisted on sticking with the GPS program. But if Parkinson had any doubts, they never showed. "That's what Brad was superb at," Green says. "People would take a club and start to hit him, and they'd always miss."

There is still disagreement over who should get the most credit for GPS, Roger Easton and the Navy or Brad Parkinson and the Air Force. Over the years, some of Easton's Timation colleagues have accused Parkinson of inflating the importance of the Air Force-only 1973 "lonely halls" meeting as the birthplace of GPS. They point to another meeting held that same Labor Day weekend, at a motel room in Virginia, at which Easton and Timation personnel hashed out details with Parkinson and his team. It was here, they claim, that Parkinson was "offered" Timation as a replacement for the rejected 621B. They maintain that recognizing the importance of this colloquy would require a reassessment of how vital certain aspects of Timation were for what became GPS. Ron Beard, Easton's second-in-command at Timation, still pointedly wonders why Parkinson and his California-based team would fly all the way to Washington just to meet with themselves.

Parkinson, who was around the same age in 1973 as Easton was when he supervised Minitrack, had a temperament and background probably better suited for the combination of gall and charm required to navigate the political obstacles that beset GPS. "It was scientists doing their things, coming up with neat stuff," Len Jacobson says of NRL and Timation. "You never got the feeling they had the political clout the Air Force and the office of the secretary of defense had on this thing."

Unlike the heavily credentialed Parkinson, who had combat experience to back up his book smarts, Easton had managed to get as far as he had with no advanced degree. Easton was beloved by his staff ("one of the cleverest people I've ever met," says Pete Wilhelm, the Timation team member), but Parkinson inspired more complex loyalties. "He was very single-minded in his pursuit," Green says. "He did well with his boss, and he did well with people that worked for him. He didn't do well with his contemporaries, because they would always try to say something and he would be so damn smart he would waltz right around them, which would then irritate them." Parkinson, who had reached the rank of colonel after fifteen years of service—about five years fewer than the typical Air Force career track—was surprised he had been promoted at such a young age, "because I tend to be outspoken," he says.

The biggest difference between Easton and Parkinson was not their personalities, or the precise structure of their ideal global navigational satellite system. It was what most animated their enthusiasm. Easton thrilled to the possibility of time transfer, and the possibility of updating the chronometric tradition. Parkinson wanted pinpoint weapons delivery, a priority made abundantly clear by a sign he posted on the wall of his office:

> The mission of this Program Office is to
> - Drop 5 bombs in the same hole
> - and build a cheap set that navigates
> - and don't you forget it

"I had great sensitivity to the fact that everything we were doing really related to the warrior," he explained. "We were trying to put together a system that would enhance and revolutionize warfare. The model that we had—'Drop five bombs in the same hole'—meant, don't forget the end product of what we were trying to do here."

From 1973 to 1978, Parkinson and the GPS Joint Program Office headquartered at Los Angeles Air Force Base, along with some private contractors, built GPS from the ground up. In June 1977, after four years spent constructing satellites, writing software, building custom computers, and designing GPS receivers, the first GPS test satellite launched from Vandenberg Air Force Base in California. The satellite, called NTS-2, spent nearly a month in a temporary orbit before being guided into its final position. Parkinson's team challenged the three companies building the receivers—Texas Instruments, Magnavox, and Collins—to be the first to pick up the signal from 20,000 kilometers away.

In Cedar Rapids, Iowa, a young engineer named Dave Van Dusseldorp, just a few years out of college, sat on the roof of Collins's headquarters, staring up at the starry sky. It was a warm summer night. Two wires next to him snaked across the roof and through an open window on the floor below. One was a simple antenna he pointed at the sky, attached to some enormously complex equipment inside. The other was attached to a telephone, in case his wife, who was eight months pregnant, needed to reach him. "The urge was just overwhelming, as you pointed the antenna, to look up at the sky and think you could see the thing," he remembers. Six minutes after the GPS signal went out, he heard the shouts and cheers filter through the window. The Collins team had found the signal. It was faint, but it was there. Van Dusseldorp had become the first person on earth to receive the GPS signal. Van Dusseldorp crawled through the window and met the four or five of his smiling colleagues who were part of the test. They were staring at oscilloscopes, their faces tinted green from the lines on the display. "It was a thrill for us engineers," he says. "We just had oscilloscope displays. Somebody could've walked in and they wouldn't have known the difference. But we knew—there it is, there's really a satellite up there and we're tracking it! I'll be darned."

Less than a year later, on February 22, 1978, the first operational GPS was launched. GPS was embryonic, but it was alive and ready to serve. Now came the real tests. For a few hours each day, the satellites were in position for coverage in the southwest United States. At the Army's vast Yuma Proving Ground in Arizona, large orange Xs were painted in different places around the desert landscape. Mel Birnbaum, an Air Force major and whip-cracking engineer who oversaw the data processing aspects of GPS, reviewed the tests. A bookish-looking man with a receding hairline and large glasses, Birnbaum had a ferocious attention to detail that could unnerve the other engineers. On at least one occasion, he'd kept them at work until 2 a.m. on a Friday. He was a good choice to run these exercises. Nothing would slip by him.

Birnbaum rode in the back seat of a two-seater F-4, his eyes fixed on a small glowing screen. A computer programmed with the Xs' GPS coordinates was on the underside of the aircraft, along with concrete replicas of bombs. As the F-4 approached the correct GPS location, the computer signaled the exact moment to release the bomb. For several days, Birnbaum brought fake death from above to the Sonoran Desert. The results were more than encouraging: GPS behaved beyond expectations. The presumed margin of error was 50 feet, but many of the bombs were accurate to within 10–15 feet. One day, after the F-4 had dropped its six bombs and landed, Birnbaum and others made their way to the X. Something strange had happened. There were only five holes. It took them a moment to realize that two of the bombs had landed on exactly the same spot. Parkinson's five-bombs-one-hole had always been an ideal, not a realistic goal, but this was damn close.

GPS was doing its job, but the program faced criticism, both inside and outside the military. At the time, the Air Force, as a whole, was not interested in precision, instead pursuing a doctrine of air superiority: the complete control of the skies via overwhelming force. The GPS receiver that Collins developed was a 270-pound workstation with two seats, with a price tag of over $5 million. The Air Force's Strategic Air Command had planned to purchase 600 receivers, but now cancelled the orders. In 1977, the General Accounting Office issued a report that was highly critical of the program, citing cost overruns. By the end of 1979, the Air Force was on the verge of killing the program.

Funding from the Pentagon and the Senate rescued GPS, but Parkinson was ready to move on. He was tired of trying to sell GPS to the Air Force. He was done with rhetorically asking, "Why do precision weapons help you?" and giving the obvious answer: "Because you only have to go once!" That was the crazy-making aspect of dealing with detractors. Sure, fine, whatever, you don't need another navigation aid, you know how to transport yourself to where you need to go—but what about a system that, after you get there, gets your payload where *it* needs to go? You won't miss—ergo, you won't have to endanger yourself further by turning around and dropping bombs a second time.

Parkinson's tireless advocacy for GPS had probably accomplished what that general who accosted him at the Pentagon years earlier had predicted: Parkinson had probably hindered his chances of moving further through the ranks. But that was all right; he liked to say that he never had "star fever," the burning desire to become a general someday. Parkinson decided to retire from the Air Force. A few years later, he returned to his alma mater to launch the Stanford GPS Laboratory, an incubator for creative GPS applications, particularly in the field of aviation.

"The point that I'm making," he says today, "is that the Air Force as a whole did not want GPS. They fought it while I was in, and after I left, my successor had the same problems. And this was even after it had demonstrated phenomenal, unprecedented accuracy." Except for one brief social call to visit his old friend Gaylord Green, twenty years passed before Parkinson again set foot in the GPS Joint Program Office.

Gaylord Green also left the GPS program, but he returned in 1985 to run it. GPS no longer sustained attacks from Congress or the Air Force, but Green felt the program was suffering from a kind of benign neglect. "All the things that GPS is today—we had that vision," Green says of the Parkinson-guided program in the early days. "When I came back, they had totally lost that vision. It was just sort of a satellite program. They were going through the motions of building satellites and launching them."

By early 1986, they couldn't even do that. NASA's space shuttle program was slated to carry all future GPS satellites into orbit. The space shuttle *Challenger* disaster in January ended that plan. NASA put the program on hiatus. It would be two years before the launch of another GPS satellite.

Two months after the *Challenger* disaster, the U.S. carried out Operation El Dorado Canyon, a raid on military facilities and training camps in Libya, in retaliation for the bombing of a nightclub in West Berlin that killed three people and wounded more than two hundred. Though considered a nominal success, the mission had heavy human costs: an unknown number of civilians were killed, including at least one child, and two American servicemen died. One American bomb narrowly missed the French embassy in Tripoli.

In the aftermath of El Dorado Canyon, the Pentagon directed the Air Force to design a new airborne precision weapon that could be counted on to hit a target. The Air Force decided to modify thirty-nine nuclear air-launched cruise missiles. The nuclear warheads were removed. They also augmented the missile's existing guidance system by adding a GPS receiver. Because the missiles still looked like their former selves, knowledge of the military's tinkering with them could threaten ongoing arms talks with the Soviets. Officially, these GPS-guided missiles did not exist.

One of the few people who knew about them was Buster Glosson, an Air Force officer who monitored the progress of research projects within the service, and recommended which should receive continued support. "I was very high on funding it," Glosson says of the missile program. "It was just a line item with a different name, so only a few people knew what it really was. Everybody else was only briefed on a piece of it, and had to accept that this was, so to speak, a 'black war program.'" In 1988, the Strategic Air Command, the wing of the Air Force that oversaw much of the American nuclear strike capabilities, quietly added these GPS-enabled missiles to its arsenal.

Two years later, near the end of the summer of 1990, Iraq invaded Kuwait, and the U.S. began deploying troops to Saudi Arabia. The incursion happened to coincide with a renewed effort to improve the GPS constellation. A new generation of satellites had been launched the previous year, and a few months before the invasion the Master Control Station at Schriever (then called Falcon Air Force Base) had begun to reposition the satellites for maximum worldwide coverage. At the same time, parts of the military, especially the Army, were beginning to realize the value of GPS. A "portable" military-grade GPS receiver called the Trimpack became available. In Army war games that pit one side with GPS against another without it, the GPS carriers always won. Col. Roland Ellis, the head of the Army Space Command, was a particularly enthusiastic proponent, demonstrating how GPS could do everything from sighting artillery to preventing a soldier from getting lost.

There was a growing realization that if Saddam Hussein ignored the U.N. deadline to withdraw Iraqi troops from Kuwait—January 16, 1991—triggering an armed conflict, GPS would play an important role in the war effort. The Air Force struggled to get the satellite constellation into decent shape for maximum coverage over the region. The solar panels on one satellite were failing. By December 10, they had been repositioned, just in time for the malfunction of another satellite's flywheels, which rotate the satellite so that its antenna points toward Earth. Technicians opted to put it in a permanent spin—2.5 rotations per minute—that kept it in a position with its antenna pointing toward Kuwait City. It was a risky maneuver that might disrupt the satellite's orbit and cause clock problems. The crew worked through Christmas to get it stable.

On January 16, a few hours after the deadline passed, the wanton satellite was finally deemed dependable enough for coverage in the Gulf. "And that," says Green, "is what allowed the war to start."

Chuck Horner was fond of the phrase *Insha'Allah*, "God willing." It was on the mind of the Air Force general the day in 1990 he arrived in Riyadh to assume temporary command of U.S. forces in Saudi Arabia. He had visited the country several times over the years, marveled at how much the principles of Islam shared with the Judeo-Christian tradition, and grew to love the sound of the daily calls to prayer. He was nervous about the impending conflict and his role in it as commander of the air component. Over the years, to demonstrate to his Arab friends that "you cannot trust America," Horner would note that "the once-upon-a-time capital of the last nation to put complete faith in American military might is now called Ho Chi Minh City."

Horner's time as a fighter pilot in Vietnam had left its mark. He had come to see the war as a "stupid, evil, aimless thing," for which he blamed everyone from Lyndon Johnson to the generals in Saigon. "I hated them because they asked me to take other people's lives in a manner that dishonored us both, me the killer and them the victim," he explained. "Shame on all of us."

Horner had chosen Buster Glosson, by then a brigadier general, to plan the Desert Storm air campaign. With his sleepy eyes and hangdog expression, Horner had a more laconic air than Glosson, a husky silver-haired Southerner who, in the right light, bore a resemblance to Dennis Hopper. Glosson sometimes rubbed Horner the wrong way, but Horner also appreciated his stubbornness. Glosson also took inspiration from *Insha'Allah*, inscribing the words on the inside cover of a diary he began on the eve of Desert Storm. Above it was an inspirational verse that had helped him get through the Vietnam War, which began "where there is faith there is Love" and ended "where there is God, there is no need."

Vietnam loomed as large for Glosson as it did for Horner, and for similar reasons. He had hated the prevailing preference among the military brass for the "brute force frontal attack," with body counts as markers of progress. "We never had a clue, at the unit level, what our overall effort was trying to accomplish," he recalled. "We became almost mechanical." He liked the idea that new technologies could enable a cleaner, more clinical approach to air campaigns. Like Parkinson, he grasped how GPS meant not having to go back a second time, and eliminated bad weather and cloud cover as impediments to hitting the target. "It doesn't make any difference if it's so foggy that ducks are *walking*," he liked to say by way of talking up the benefits of GPS. "The bomb still explodes. On target!"

Determined not to make the same mistakes their superiors had made in Vietnam, they decided that Desert Storm would begin by taking control of the air, with an overwhelming show of force that would hopefully shorten the conflict. It began at 2:30 a.m. local time on January 17, as two Air Force Pave Low III special operations helicopters lifted off from an airfield in Saudi Arabia. They were followed by four Army Apache attack helicopters, flying 200 feet above the Pave Lows. Shooting through the inky darkness at speeds of nearly 400 miles per hour, the entire caravan headed for its target, launching the first offensive maneuver of Desert Storm. The Pave Lows dropped glowing yellow chemlites as guides, a cascading aerial yellow brick road.

When Horner and Glosson had proposed this opening maneuver, called Project Normandy, to H. Norman Schwarzkopf, Jr., the Army general leading the Coalition troops, he was not impressed. The problem was the symbolism: the Air Force leading the Army into battle. "He went berserk," Horner says. "We had to calm him down and say, 'No, no, the first weapon used in the war will be United States Army.' The Special Operations forces could've knocked those radar sites out, but Schwarzkopf was *not* going to have Special Operations fire the opening shot in the war." Glosson remembers a more measured response from Schwarzkopf, a reassurance that the Army would be part of the "opening salvo." "I was not a neophyte, politically, so I knew what he meant," Glosson says.

If the optics were a problem, why not just have the Apaches lead the Pave Lows—or just remove the Air Force from the mission? Because the Pave Lows, unlike the Apaches, were equipped with GPS receivers. They could fly through the night, confident they were traveling on the correct path, so that the Apaches could launch their missiles precisely at the planned time. That was enough to convince Schwarzkopf, who had already bucked his branch of the service by prioritizing the air campaign and preempting the ground war. "The Army wanted to shoot from day one, and Schwarzkopf said no, we're gonna do it this way—we're going to take air power and make it so that when the Army goes to war there are few casualties," Horner says. "Schwarzkopf truly cared about the lives of soldiers. He'd been in Vietnam. He'd carried wounded guys from minefields. Innately, that guided him more than any other thought."

"They both understood, because they had the same scar tissue from Vietnam that I did," Glosson says of Horner and Schwarzkopf. "They wanted minimum loss of life. And to them and to me, that meant zero, if possible. There wasn't any number that we agreed we'd lose to do this. We wanted to lose zero."

If everything went according to Normandy's plan, the Hellfire missiles from the Apaches would destroy the radar sites just moments before they would be able to detect the approach of Air Force Ravens and stealth Nighthawks, which would then have a 20-mile-wide safe corridor for bombing runs on downtown Baghdad. Glosson was concerned that if Normandy failed and the Iraqis realized what was happening, they might take the opportunity to launch Scud missiles at Israel.

A few minutes into the journey, the Pave Lows gave their final position update to the Apaches, and then got out of their way. The Apaches split into two groups, separated by a half mile, and launched their Hellfire air-to-ground missiles at two Iraqi radar defense sites, 70 miles apart. Before the missiles had even reached their targets, the teams banked and began flying back to a remote staging area in the desert, guided once again by the Pave Lows.

GPS made other contributions in the first hours of Desert Storm. With the radar sites destroyed and the safe flying zone opened, the bombing of Baghdad began. Less than an hour later,

Coalition forces launched thirty-nine drone aircraft from two sites in the desert, using GPS to calibrate the launch positions. The drones revealed the locations of more Iraqi radar sites. A squadron of thirteen B-52s, partially navigating by GPS, flew low to destroy airfields and a landing strip. During that initial punishing air campaign, B-52s were in constant motion. A U-2 reconnaissance aircraft would find a target with a GPS-assisted radar and transmit the coordinates to a surveillance aircraft that would forward the data to the B-52. The data was entered into the navigation computer and the plane would go to its target. The crew would then use GPS to time the bomb drop.

GPS had the greatest impact in the ground offensive that began the second week. The Iraqi army had assumed the Coalition forces would be at a disadvantage in the vast, faceless desert, limited to the few major roads and highways. But thanks largely to GPS, the Coalition accomplished the first large-scale deep desert advance in the history of warfare. Knowledge of GPS coordinates allowed tanks and mechanized infantry to move quickly, cutting down on the risk of accidents and friendly fire, especially during the first forty-eight hours of the war, when bad weather caused visibility to drop to as little as five meters. Soldiers found water sources by following goat tracks and marking the spots with GPS. They used GPS to report on the presence of mines, and for positioning artillery. Meal trucks used GPS to deliver food to troops. Special Forces, disguised as civilians, included GPS coordinates with intelligence on targets. When it was over, GPS was used to map coordinates of unexploded mines—and dispose of them—over an area twice the size of the Greater London metropolitan area.

GPS was arguably the primary reason Desert Storm ended so quickly and decisively. And yet the total number of GPS units on the ground was, at most, 7,000, for a Coalition force of half a million troops. Fewer than 900 were considered military-grade. The rest were early commercial models. Although the actual saturation level of GPS in Desert Storm was low, its impact as a sort of global meme was high. As Thomas Moorman, leader of the Air Force Space Command, put it, GPS "came of age in the Arabian peninsula." It was certainly the moment the military finally and firmly embraced it, and also the moment it entered the public's consciousness. "GPS really came before the public in Desert Storm," Horner says. It was not uncommon for media reports to hail GPS as the war's "unsung hero," although other new technologies, such as stealth aircraft, were at least as important.

GPS dominated the discourse because it stood for a new kind of warfare, an evolution of Westmoreland's automated battlefield. "The Chinese watched the Gulf War with great interest, because they saw a Soviet-trained client force decimated by space-based technology," says Scott Pace, the George Washington University professor. Russia's most prominent military scientist noted that Desert Storm showed that terms like "front lines" and "flanks," and the idea that winning a war means occupying enemy territory, were no longer relevant. Indeed, one of the reasons the Air Force finally embraced GPS, aside from its obvious utility, was that it dovetailed with a phrase the Air Force had begun to use as a mission statement: "Global Reach, Global Power."

This was the technological legacy of Desert Storm: the idea that warfare was not bound by geography. Even if GPS use in Desert Storm was barely automated, it held out the promise of weapons whose precision was derived from GPS, rather than merely aided by it. That mode of waging war actually debuted in Desert Storm, but it was done in secret.

History considers January 17, 1991, the first day of Operation Desert Storm. In fact, it began a day earlier, when Air Force engineers were still struggling to ensure that the satellites were providing robust coverage over the Gulf region. At 3 a.m. on January 16, three hours after the deadline for withdrawal, fifty-seven airmen at Barksdale Air Force Base in Louisiana prepared to climb into seven hulking B-52 Stratofortress bombers. They took off in heavy rain. Because of the 244-ton payload carried by each aircraft, like an insect's bulging egg sac, they required nearly two miles of runway to get off the ground.

They were executing Operation Secret Squirrel, which the airmen had trained for in private—a "black op" which they were prohibited from mentioning to anyone. President George H. W. Bush had personally signed off on it. Horner and Glosson had approved it. Very few others were aware of it. The B-52s were carrying CALCMs, those nuclear cruise missiles that had been denuked and installed with GPS receivers, which officially did not exist. To conceal their mission and avoid attracting attention, the planes flew low and slow across the Atlantic, finally reaching Saudi airspace after 8 a.m. local time the next day. At around 8:30, six hours after Normandy began the war, the crew launched their missiles at Iraqi targets hundreds of miles away. Some were programmed to travel in a straight line, others more elliptically, so that all the missiles would reach their targets, which included a power plant in Mosul and a communications center in Basra, at the same moment.

Aided by GPS, most of the missiles behaved flawlessly—one, whose bull's eye was a telephone pole, snapped it clean in half. Two missed entirely; one was lost; another failed to detonate. The crews then turned their aircraft around and flew back to Louisiana. They had spent thirty-four straight hours in the air and covered 14,000 miles, the largest combat mission ever flown.

The U.S. did not confirm that the mission had happened until exactly one year later, when it was mentioned, with no fanfare, at a Department of Defense press conference. The oddest aspect of secrecy is that all that effort had been exerted for targets that were not considered essential, in the way Normandy's were. Why fly thousands of miles to hit a barely tactical telephone pole from hundreds of miles away? At the time, Glosson said Squirrel mattered most as a "political statement." But a statement about what, and directed toward whom?

"I meant that, let's be honest, you go to war for political reasons," Glosson says today. "It is a profound political statement for a nation-state to take off within its own boundaries and deliver a weapon as precise as [this one] was. It showed that if we wanted to expend that many resources, we could make a significant impact halfway around the world. And nobody could do anything about it. We didn't need lands, we didn't need bases, we didn't need anything." Global reach, global power. Brought to you by GPS.

When the war was over, the military began to do for bombs what Squirrel had demonstrated with long-range missiles: make them smart. Ron Yates, head of the Air Force Systems Command, the service's research and development arm, led the way. Yates, firmly on the operational side of the Air Force, didn't really register GPS until around 1986 ("GPS was a space system, and I wasn't concerned with anything that didn't have to do with fighters"). When he learned more about it, his thoughts were similar to those of fellow vets Parkinson, Glosson, and Horner. "I was part of dropping 'dumb' bombs in Vietnam," he says. "When people are shooting at you, guys start dropping bombs from really high. Instead of dropping them from 4,000 feet, they'd be dropping the freaking bombs from 10,000 or 12,000 feet, at 650 knots, and never hit anything."

Yates supervised the construction of six bombs with integrated GPS receivers at Elgin Air Force Base, on the far western end of the Florida panhandle. To test them, Yates organized trials similar to the exercises the original GPS team had conducted in Arizona. Yates's tests were more dangerous, since they removed the human factor. In lieu of an attentive Mel Birnbaum using GPS to time a bomb's release over the desert, these bombs would guide themselves. Once the testers chose a target on the bombing range, they double-checked the coordinates—to make sure they wouldn't accidentally rain fire down on nearby Fort Walton Beach—and transmitted them to a pilot, who entered them manually on a bomb's receiver. "We had somebody else up there to make sure that yea, verily, the pilot hadn't screwed up and targeted the beach," Yates says.

This was the general scenario Yates envisioned for using this kind of bomb in actual warfare: someone on the ground radioing coordinates to someone in the air. For the test, the target was a crude shack 14 miles away from the launch point. The bomb completed the journey successfully. Yates was elated.

These experimental bombs would eventually evolve into the Joint Direct Attack Munition (JDAM), a kit that converts a conventional dumb bomb into a GPS-guided smart one. The JDAM officially entered the annals of warfare on March 24, 1999, when two B-2 stealth bombers took off from Whiteman Air Force Base in Missouri. They were part of Operation Allied Force, NATO's bombing of Yugoslavia in response to Serbian aggression in Kosovo. They flew for over thirteen hours before reaching Yugoslav airspace. Virtually invisible and impervious to radar, the pilots took their time, spending several hours scanning targets and letting the B-2s' targeting computers do their job. The system created extremely detailed radar maps, which were then merged with data from the B-2s' onboard GPS receivers. The bombardiers used trackballs to select targets from the maps, dropping sixteen JDAM-enhanced bombs.

But there is a corollary to being able to drop five bombs in the same hole: what if you have the wrong hole? A few weeks into the 1999 NATO campaign, a B-2 sent a JDAM to a military target in Belgrade. It found the target flawlessly, but there was a mistake: the bomb hit the Chinese embassy, and three journalists inside were killed. The problem was not the JDAM or the person who entered the coordinates. The CIA, which had provided the coordinates, had used an outdated map.

Even as GPS weapons eliminate human error, it reasserts itself in new ways. In the last days of 2001, as fighting intensified in

Afghanistan, an Air Force combat controller in a command post surveyed the GPS coordinates of a Taliban outpost and prepared to transmit them to a bomber in the air. His GPS receiver's battery was dying, so he paused to replace it. What he did not know, or forgot, is that the machine was programmed to show its present coordinates when turned on. The coordinates he transmitted were his own. Instead of destroying an enemy outpost, he had summoned a 2,000-pound JDAM which obliterated the post, killing three Green Berets and many more Afghanis who were part of the Northern Alliance, but sparing the life of the country's future president, Hamid Karzai.

GPS is also entrenched in the military on the ground level, though here military contractors are losing their foothold. A recent survey of GPS military use among troops in Iraq and Afghanistan found the "most prevalent unit" for accessing GPS was the iPhone. Even though mobile phones cannot access a special encrypted hyperaccurate GPS signal reserved for the military, they augment their position readings with data from other satellite systems, Wi-Fi, cellular, and Bluetooth networks. Survey responders consider the standard military-grade receiver, the DAGR ("dagger"), manufactured by Rockwell Collins, to be essential for environments where the GPS signal is jammed, and for interfacing with other military equipment that uses GPS; they almost never use it as a handheld device for quick position readings. One commenter claimed that the DAGR model shipped to troops in 2005 "stopped being functional . . . almost the day it was delivered."

It took nearly two decades and a decisive global crisis for the military to see the possibilities of GPS. The rest of the world took to GPS much more quickly, and immediately began to improvise extraordinary new uses for it.

CHAPTER FOUR

Ranging the Perfect Beet

Among all the varieties of the cultivated beet plant—red, golden, chard, blood turnip, the Chioggia with its snappy bull's eye pattern—the sugar beet really puts the "vulgar" in *Beta vulgaris*. These unattractive lumps of vegetable matter are inedible and difficult to grow. But they conceal an inner secret. They are extremely adept at absorbing minerals and moisture from the soil and transforming them into sugar.

Out on Colorado's Front Range, where the Rockies rise from the Great Plains like a bouncer behind a snowy velvet rope, the sugar beet was once practically a way of life. The Colorado Gold Rush of the mid-nineteenth century lured prospectors to the region. Most had no luck as miners, but some decided to stick around and build a life. A few even tried to farm, a vocation only somewhat less foolhardy than panning for gold. The Front Range was not quite a desert, but it felt like one. The famous newspaperman Horace Greeley called it a "land of starvation."

Those who tried planting sugar beets made a startling discovery: the Front Range had been created by a beet-loving god. The clay and silt in the soil provided minerals sugar beets craved, and the sparse rainfall kept those minerals from washing away. The hot summers maximized the beets' internal sugar production, and the cool autumns slowed the process at its peak. Even with these prime conditions, growing sugar beets required finesse and a larger investment than other crops. Once the plants sprouted, a team of thinners walked the rows, yanking plants to limit competition for resources. At harvest time, the farmer dragged an implement that loosened the beet's central root, while laborers followed, carefully lifting it from the soil. Another team sliced off the tops, careful not to leave any leaves attached (the untrimmed leaves fermented, potentially spoiling a whole pile of stored beets), but careful also not to cut off any part of the root, which was like throwing money away.

A half century after the first Front Range sugar beets came to market, the industry exploded. At processing plants that dotted the area, beets were washed, sliced, and boiled to extract a syrup, which was then filtered, carbonated, combined with lime and sulfurous acid, evaporated, crystallized, and centrifuged to produce white sugar. Sugar factories had a catalytic effect on sleepy Front Range towns, creating thousands of jobs. The capital injection allowed municipalities to build a modern infrastructure. As early as the 1880s, Fort Collins, county seat for one of the biggest beet-producing regions in the state, had a modern water delivery system and one of the first electricity plants in the West. For the next forty years, Fort Collins, according to one local historian, enjoyed a prosperity based on "the cultivation of beets and the feeding of lambs." Adjusted for inflation, the Great Western Sugar Company, America's largest producer of beet sugar, paid local farmers about $140 million annually for their crops. Research on sugar beet farming led to improvements in irrigation and pest control that improved agricultural practices in this land of starvation. Sugar beets grew the Front Range as much as the Front Range grew sugar beets.

Gone are the days when the Great Western Sugar Company, America's largest producer of beet sugar, powered its Front Range factories with coal from its own mines, extracted lime from its own quarries, and transported products on its own private railroad. Today, the Front Range urban corridor, snaking through 200 miles of Interstate 25—from Pueblo, Colorado, to Cheyenne, Wyoming, with Denver in the middle—is one of America's fastest growing "megaregions." The Front Range's biggest industries—information processing, high-tech, aerospace, defense—all come together in GPS, which is "manufactured" near Colorado Springs, at Schriever Air Force Base. Two hours up I-25, as far north of Denver as Colorado Springs is south, sugar beets are still part of the culture and economy of Fort Collins. If Colorado Springs embodies the new Front Range economy, postindustrialism built on the humus of the region's prior agrarian boom, Fort Collins is where that loamy past survives in the present by harnessing the future.

The world still needs beet sugar—for food additives, livestock feed, and even biofuel—but it remains a demanding crop. Everything that goes into growing a beet—planting, nurturing, harvesting—now requires the use of GPS, augmented by other satellites, including some belonging to the Kremlin and others built with the express purpose of helping GPS land airplanes.

GPS is such a torrential global presence that it's easy to forget that the gatekeepers once took great pains to ensure that all we'd get is a trickle. GPS can not only find a beet, it can find the exact spot to place a beet seed. Not long ago, it wasn't even guaranteed to be accurate enough to locate the beet *field*.

From the earliest days of the GPS Joint Program Office, everyone involved understood that their creation would have some sort of universally accessible component. "We wanted civilians to use it," says Gaylord Green. But given that this was a military project, the Department of Defense had two valid concerns. First, why should it fund a program that would give military and nonmilitary users, including enemies of the United States and its allies, equal accuracy? And if the system was this easy to use, what could be done to protect the system from sabotage?

Their answer to the first question was to allow civilians access to GPS in a way that did not exploit its full potential. The Air Force assumed that the largest nonmilitary user segment for GPS would be amateur sailors. They were already using either the current version of the land-based LORAN technology, Loran-C, if they wanted the quickest reading, or Transit, if they wanted the most accurate reading possible and didn't mind waiting for the satellites to come into view. The Air Force polled 500 of these users, asking what kind of accuracy they would expect from a satellite-navigation system. The most common response was a margin of error of 500 meters—no surprise, since that was the accuracy of LORAN. This was good news for the GPS team: 500 meters, they could live with.

The team decided that the GPS signal would come in two flavors: the coarse acquisition (C/A) code and the precision (P) code. The P code would be longer than the C/A code, and therefore able to provide more detailed data to the receiver, which would result in better accuracy. It would also be more complex, its informational content buried in a greater amount of pseudorandom noise, which increased its security.

GPS signals would broadcast on two channels: L1 would carry both codes modulated onto one signal, though only military receivers would know how to dis-entwine the P code; L2 would carry only the P code.

Placing the P code on two channels added another layer of security. If an enemy managed to jam one channel, a military user could switch to the other. It also further increased accuracy by correcting for any delays caused by the signals' encounter with the ionosphere.* Experiments suggested that the P code

* If a GPS receiver can process both signals simultaneously, its programming allows it to detect—and correct for—any ionospheric delay problems.

would provide positioning accurate to within 10 meters. The C/A code's margin of error would be about ten times larger—not great, but five times better than mariners expected from a navigation system.

The tidy border between military and civilian GPS began to blur almost immediately after the launch of the first operational GPS satellite in 1978. The following year, Magnavox, the company that designed the spread-spectrum coding used by GPS, introduced the first GPS receiver for the nonmilitary user. The Z-set wasn't very portable (a receiver and a display unit) or cheap ($15,000), but tests showed that it was unnervingly accurate, barely less so than military receivers, despite its lack of P code access. It also beat the military receivers into war. In what was probably the first operational use of GPS by the armed forces, a Z-set was on board one of the helicopters used in Operation Eagle Claw, the doomed 1980 attempt to free hostages held at the U.S. embassy in Tehran.

As other civilian receivers began to appear, the GPS program conducted its own tests. The results were vexing. "We thought the civilian receivers would give you about 100 meters, and they ended up giving us 30," Len Jacobson says. "And that's too good to give away to the enemy. So we came up with a scheme to screw up the signal for civilians."

The plan to make civilian GPS worse was led by Mel Birnbaum, the Air Force software engineer who had worked so hard to make GPS perfect, and dropped the faux bombs in the 1977 test that demonstrated it was. Implementing the idea would require a new generation of satellites, still a few years away.

Launched in 1938 to market audio oscillators, Hewlett-Packard was a kind of Platonic ideal of a Silicon Valley startup, right down to its origins in a Palo Alto garage. Within the tech industry, HP became known as much for its innovative products as for its adherence to an oddball (now familiar) corporate culture it dubbed the "HP way": egalitarian ideals, empowerment of employees, a reverence for individual ingenuity. By the 1950s, HP had instituted flextime policies, casual-dress Fridays with end-of-workday "beer busts," and workspaces where only H and P had their own private offices, which they mostly just used for meetings.

HP hit its stride in that period, becoming a juggernaut in the instrumentation industry, designing high-quality products like voltmeters and frequency counters. HP solidified its grip on the instrument market in the early 1970s by integrating digital controls into multipurpose, intelligent, and increasingly automated analyzers, testers, and calibrators for microwave networks, chemicals, and digital systems. The company was exploding—between 1969 and 1977, its annual revenues quadrupled as the size of its workforce doubled. HP's 35,000 employees made it one of America's ten largest manufacturers. It was a remarkably fecund period for HP, as scores of those 35,000, with tacit approval from the very top and active approval from middle managers, went looking for the next big thing.

In 1973, an HP engineer named Ralph Eschenbach transferred to the company's corporate labs after five years in a division that made semiconductors. Corporate labs was a skunkworks R & D division, and employees were expected to devote about one-third of their time to ideas that might lead to an innovative product somewhere down the line. "It didn't need to be officially funded and sanctioned," Eschenbach says. "You just went and did it on your own." A year into his stint there, Eschenbach read about GPS in a trade magazine. An avid sailor, he immediately understood its utility. "I said, 'Boy, wouldn't it be fun to build a product like that!'" he recalls.

Toward the end of 1976, Eschenbach drew up plans for a GPS receiver. He had studied some of the receiver designs made by companies like Rockwell Collins, Magnavox, and Texas Instruments, discovering what looked like a persistent flaw. "What they did with all those millions of dollars is they put one guy on the antenna, another on the digital, some guy on the memory, another on the display," he says. "Everyone optimized what they were doing, so you got a lot of optimized subparts, but not an optimized whole. I had the whole design. From antenna, to digital, to software, I had to get the job done." Eschenbach also analyzed the GPS signal structure, eliminating parts that seemed superfluous to the needs of commercial users, and streamlining the processing. Instead of having the system perform calculations once a second, which taxed the era's microprocessors, Eschenbach designed a quadratic equation that estimated the satellites' positions for five-minute periods.

Eschenbach and a small team built a prototype GPS receiver containing an eight-bit processor, to handle the satellite tracking algorithms, connected to an HP 9825, the company's newly released desktop computer. They picked up signals from the GPS test satellites that were already circling Earth, staring at the green-on-black screen of an oscilloscope to judge how well the receiver performed. "I just turned the receiver on and worked with it, and the head technicians all hovered around," he says. "It was an exciting time, sort of like Watson and Bell, you know?"

Sometime in 1981, they decided to try out their contraption. They rented an RV and set up the GPS receiver on the kitchen table, connecting it to an antenna on the roof. Eschenbach attached the HP computer to a chartplotter, a device used in navigation to translate navigational information onto a map by plotting points within an x/y coordinate system. The chartplotter contained an ink pen that would place little dots on the map as the receiver processed the GPS information.

They took their mobile GPS unit onto Interstate 280, the freeway that now connects San Francisco with Silicon Valley. As they were driving, the dots began to appear, running smoothly along the thick line on the map that denoted I-280, curving where it curved and straightening where it straightened. "It was fascinating, just thrilling," Eschenbach says.

But then the dots went rogue, suddenly not cleaving to that colored line. Eschenbach was stricken. All that work . . . And then it became clear what the problem was: I-280 was still under construction. It had only been extended to the South Bay a few years previously, a change not noted on the paper map they were using. The receiver was working perfectly. It was the map that was wrong.

Hewlett and Packard were impressed by what Ralph Eschenbach had accomplished. But despite Eschenbach's impassioned lobbying, and after about $800,000 spent developing the receiver, the company passed on taking it to market. By the early 1980s, the company had decided to transition from the instrumentation market, which it dominated, to make a run at the $40 billion computer industry, of which HP controlled just 5 percent. The company's founders determined that the next frontier in consumer tech was the home computer, not the personal navigation system.

It was a logical conclusion. Hewlett and Packard saw two possible markets for GPS. The automotive industry might conceivably be interested in installing GPS navigation systems in cars, but the case was hard to make when the incomplete constellation meant sporadic coverage. That left the marine market—and here the problem was more cultural than technological. "HP's sales force in those days was like IBM's," Eschenbach says. "They wore button-down shirts, blue coats, and ties, and they couldn't imagine their salespeople going down to the docks and dealing with fishermen." If there was a flaw to HP's reasoning, it was its failure to imagine that people who dressed like their sales force might themselves be a potential market. Might not the white-collar worker want GPS someday?

Charlie Trimble thought so. For him, Eschenbach's experiment, which proved beyond a doubt that the civilian GPS code was capable of providing readings nearly as precise as the military's, made GPS—available free to everyone—an extremely undervalued commodity. He thought of another communication method, the long-distance phone call, and how quickly its cost was dropping, on its way to pennies-per-minute. It was though the completely gratis GPS signal had already leapfrogged the dial tone. There was this amazing heartbeat from the sky just waiting for people to exploit its possibilities.

Trimble had joined HP in 1964, fresh out of Caltech. He had shelved plans to go to Harvard Business School, lured by an offer to manage his own project. The company wanted him to work on a device for the biomedical industry that would extract information from brainwaves—"fundamentally a problem of pulling signals out of noise," he says, an idea that would occupy him in a different form many years later. Trimble left HP to form his own company in 1978. He took with him, at a cost of $50,000, the rights to a cancelled HP plan to market a Loran-C receiver, targeting the amateur marine market.

By 1982, Trimble Navigation was a modestly successful company, but Charlie was getting restless. He was intrigued by the work Eschenbach had done with his GPS receiver, and he had ideas about how to make it work as a commercial product. His one major reservation was whether the GPS program might dry up and blow away. He sought the advice of Brad Parkinson, who told Trimble that if the project had survived the 1970s, it was likely here to stay. As it turned out, one of the major Cold War geopolitical crises of the 1980s ensured that it was.

On the final day of August 1983, a Korean Air Lines Boeing 747 bound for Seoul left New York and flew to Anchorage for the first leg of its journey. After departing from Alaska, sometime during the early morning of September 1, the plane was shot out of the sky over the Sea of Japan by two missiles fired by a Soviet pilot, just off the Russian island of Sakhalin. The accident claimed the lives of 269 people, including Larry McDonald, a congressman from Georgia.

The plane had drifted 300 miles off course, and had been flying in Soviet airspace for two minutes prior to the attack. Most likely, someone in the cockpit had punched the wrong coordinates into the inertial navigation computer* while in the air. Far out over the ocean, the plane was beyond the reach of any air traffic controller. The only contact it could have with anyone on land was a high-frequency radio link through which messages were relayed, but there would be no way for any controller to note the plane's drifting course.

The politics of navigation quickly became a hot topic within the U.S. government. Four days after the incident, President Reagan issued an executive order directing the Department of Defense to release classified descriptions of the GPS signal structure. "World opinion is united in its determination that this awful tragedy not be repeated," a White House statement read. "As a contribution to the achievement of this objective, the President has determined that the United States is prepared to make available to civilian aircraft the facilities of its Global Positioning System when it becomes operational in 1988." That was the year GPS was scheduled to have its full constellation of twenty-four satellites, when worldwide twenty-four-hour coverage would be available. There were calls in Congress to move that date to 1985.

On the surface, there wasn't much to Reagan's directive. Details of the GPS signal were already readily available to anyone who wanted to build a receiver that could read the C/A code. But there was always the possibility that the military could, if it wanted to, change the GPS signal enough to make current receivers unworkable, impose a user fee (the Pentagon had been kicking around the idea of instituting an access charge for anyone

* An inertial navigation system is one that uses sensors to compute positioning, velocity, and direction, based on dead reckoning.

without "official U.S. user" status), or even cancel the program outright. The administration's directive was an open letter from the executive branch of government, informing the military that GPS was no longer theirs alone.

The Pentagon's response was to reassert its control over the part of GPS that still belonged to the military. If this tragic accident inspired more people out there in the world to use the C/A code, the military could at least make it even harder to use—or disrupt the P code. The solution was a layer of encryption modulated on top of the P code. To process the P code, a GPS receiver would not only require knowledge of how to disentangle the code from its cocoon of pseudorandomness, it would also need a decryption key to unlock it.

Ralph Eschenbach was still working for HP when Charlie Trimble purchased the rights to the GPS project he began. Eschenbach began moonlighting as a Trimble consultant, coming into the office once a week after work for five-hour pizza-fueled consulting sessions with the three engineers Trimble had hired. Far from being bitter about ceding control of his creation, he was thrilled to see it get a second life. "I basically used his engineers instead of my technicians to start from scratch," Eschenbach says. "Rarely do you get this opportunity in engineering—to learn from things you did once, and redo it all again." Under Trimble's supervision, they determined ways to make Eschenbach's original receiver leaner and more robust, streamlining the algorithms Eschenbach had devised, and figuring out ways to make the data processing even simpler—"one more iteration of perfection," Eschenbach says. After almost two years of consulting, Eschenbach quit his job at HP to become Trimble Navigation's vice president of engineering.

The third rung of the Trimble triumvirate took a more circuitous route to get to the company. One day in 1981, Javad

Ashjaee had arrived at his office at Aryamehr University, in Tehran, and heard the unsettling news that a fellow professor had been murdered. Like Ashjaee, this man was one of four faculty members serving on the university's senate. A young chair of the Aryamehr's computer science department, and the driving force behind the school's first microprocessor lab, Ashjaee was also an outspoken critic of the Ayatollah Khomeini and the chilling effect the Islamic Revolution was having on academic freedom. He had assumed that he was under surveillance by the Revolutionary Guards, the new government's internal security force, but the death of his colleague confirmed that his own life was in danger. He decided he had no choice but to flee the country immediately, leaving behind his wife, two young daughters, and a comfortable middle-class existence.

Ashjaee had lived in the U.S. previously while working toward advanced degrees in math and electrical engineering at the University of Iowa, so he decided he would fly to Zurich, and then, hopefully, to the U.S. At the airport, Ashjaee recognized other university colleagues trying to escape Iran—he looked straight ahead to avoid making eye contact with them. In the customs line, Ashjaee pleaded with a guard who said his visa was missing a necessary stamp. The guard stared at Ashjaee for a few seconds, taking his measure, and then waved him through.

Ashjaee boarded the jet to Zurich, sat back in the seat and caught his breath, and looked out the window. Police cars were speeding across the tarmac toward the plane. This is it, he figured, resigned to his fate. But the cars sped past the plane, apparently in search of someone else. Ashjaee took a last look outside, fastened his seatbelt, and waited for takeoff.

Ashjaee settled in the Bay Area, hoping to snag a tech job and eventually bring his family to the U.S. He'd been there a week when he answered an ad seeking an engineer at Trimble Navigation. Charlie Trimble called him the next day, and that afternoon Ashjaee had a desk and a directive to work on the system software for Trimble's products. A few months later, Trimble mentioned to Ashjaee that he was thinking of abandoning Loran-C and switching to GPS products. Ashjaee had never heard of GPS, but he was transfixed by Charlie's description. The mere idea that satellites 20,000 kilometers away could provide precise positioning moved Ashjaee. "It had everything under the sun that relates to electronics, except for high voltage," he recalls. "When we started, it was so fascinating that I didn't want to sleep."

To navigate the difficult process of obtaining a work visa, Ashjaee took a day job with a tech company known for its ability to streamline the process for foreign nationals. When 5 p.m. rolled around, he'd jump in his car and drive to the Trimble office, working late into the night. In a struggling company with barely a dozen employees, Ashjaee's enthusiasm and work ethic helped him form a special bond with his boss. Charlie Trimble's severe nearsightedness—he could often be spotted reading documents with his glasses pulled down, the paper inches away from his eyes—prevented him from driving at night, so Ashjaee would often ferry his boss home. They would continue planning GPS systems during the drive. "I referred to him as my mentor," Ashjaee says, "and he referred to me as a jewel he'd found."

Ashjaee spent most of his first year at Trimble writing the software and firmware for the company's first GPS receiver. "Javad was a very industrious, hard worker," Eschenbach says. "He was a blunt, brute-force twenty-five-hours-a-day beast. I didn't consider him that innovative or that creative, but he was very dedicated. He would overpower a problem simply by doing it forty-eight times."

Trimble released its first commercial GPS receiver in 1984. The first market to embrace it ignored the navigational and positional capabilities of GPS. They wanted it for its clock, a function that particularly benefited from Ashjaee's coding work. Timing centers around the world began to use GPS time to regulate their clocks. It only takes one GPS satellite to calibrate a clock—as opposed to four for a position fix—so the sparseness of the constellation wasn't an issue. Next came the oil industry. Trimble receivers were accurate to within a few feet, a small enough margin of error to calibrate positioning systems on offshore oil rigs. And if Trimble engineers could find a way to make their receivers even more accurate—say, down to the centimeter level—they would have a chance at cracking the surveying market.

By 1985, Trimble was at work on a compact model aimed at the company's old standby market, marine users. The model they devised, weighing 35 pounds and selling for $20,000, got an unexpected high-profile endorsement from conservative icon William F. Buckley. "Navigation used to be deadly serious stuff, and it still can be," he wrote in a *New York Times Magazine* feature, describing the experience of being lost at sea. "But I can report today that the problem is about to end." GPS, he promised, "would change everything." Buckley was one of the first people outside the immediate GPS community to publicly wax rhapsodic about the technology. His slightly exaggerated vision included GPS enabling blind aerial landings and helicopters flying sideways between trees at night, helping cops, cartographers, and first responders, and car systems that would "describe the sights and historical curios en route."

But he was most excited about the immediate effect GPS would have on boating. "Think big," he urged sailors. "Schedule a rendezvous with another yacht somewhere in the Bay of Fundy on a foggy day—the equivalent of hitting a bullet with a bullet. GPS will make that possible."

Buckley wrote that he was about to sail from Hawai'i to New Guinea—a trip, he noted, that was mostly out of reach of Loran-C, and thus navigable only by the stars. Trimble contacted Buckley and offered him the use of an experimental version of the GPS receiver they were building. Buckley visited the office, which he described as "right out of Walt Disney's seven dwarfs' workshop. It consisted of only a few sheds, a few dozen informal, busy and terribly bright people, computers, and warehouse, microchips, and high enthusiasm."

The day before Buckley arrived, Ashjaee had driven around San Francisco with a Trimble 4000A in the trunk of his car recording the car's position every ten seconds. He showed Buckley the results ("his eyes were shining bright," Buckley wrote): a sheet of connected dots that exactly mimicked a street map of the city. "Whether they travel on the ocean, land or in the air, all the travelers in the world will smile when GPS is finally, completely here," Buckley wrote. "It would be fine to come up with a spiritual counterpart to the GPS, so one could plot a perfect course through life—exactly knowing and achieving ideal positions respecting God, country, family, friends. But such fixes will remain inscrutable, though precious little else any longer is."

"It was a little emotional," Ashjaee says of his meeting with Buckley, whom, he felt, saw in the Iranian exile "somebody capable of doing things he couldn't do in his own country, and then coming to America and doing them."

By now, Ashjaee was settled into his new life. After more than a year, his family had finally been able to leave Iran to join him. "It took them 484 days to get out—forty days more than the American hostages," he says. But in 1986, Ashjaee was suddenly gone from the company. "It was not my decision," he says today. "There was no alternative. This is a dark part of my life, and Charlie's." All Charlie will say is that Ashjaee "left under less than happy terms."

"I was in charge of engineering then, and we had to get more engineers on the project," Eschenbach says. "We just had too many things we needed to do, and Javad wanted to run it all himself. He got to the point where he was secretive about the code he was working on, and he wouldn't tell anybody else about it."

"Javad wanted his own company," one early Trimble employee

says. "I guess nobody's quite willing to say that Javad basically designed his own product while he was still an employee of Trimble. Honestly, that's what happened."

"Trimble wasn't very fair to Javad," says Donald Mitchell, who at the time was an engineer at Datum, a precision timing company that partnered with Trimble on its early receivers. He was impressed by Ashjaee's technical skill, as well as by his honesty in pointing out and tweaking problems with the receiver. "I think management really gave him a raw deal," he said. "Mostly, the guys that were HP made out pretty good. Javad wasn't one of those HP guys."

After leaving Trimble Navigation, Ashjaee founded his own precision GPS company, Ashtech, which grew to be a formidable competitor of Trimble. Ashtech made history in 1990 when the company announced it had signed an agreement with the Russian space agency to build the first receiver to use GPS and GLONASS, the Soviet system that had begun launching satellites in 1982, four years after the first GPS satellite. At the 1990 Institute of Navigation expo, Ashtech flew a banner depicting the American flag and the Soviet hammer and sickle. Gaylord Green denied Ashjaee's request to pose for a photo under the poster.

When NASA put the space shuttle program on hold following the *Challenger* disaster, halting the launch of GPS satellites for the foreseeable future, the major military contractors working on GPS also stopped their projects. The slowdown was a boon for burgeoning commercial GPS ventures, which now had time to catch up. "That was the best thing that ever happened to us," says Ed Tuck.

A lemons-from-lemonade outlook came naturally to Ed Tuck. One day in 1985, he was flying a small plane along the Northern California coast. Stratus clouds blanketed the sky, limiting his visibility. He'd heard about GPS and the expensive receivers coming on the market. He started thinking about how great it would be if, right now, he had a portable GPS receiver for aircraft that he could use for guidance as he flew through the clouds. Even better, what if you could take GPS with you anywhere?

Tuck had in mind an ideal customer he called Bubba. The thing about Bubba was, he didn't like to ask for directions. Tuck was born in Memphis and moved to Springfield, Missouri, at the age of fifteen, when his father became the chief mechanical officer for the St. Louis–San Francisco Railroad. "Bubba is probably more south than Missouri," he says. "But I knew the culture, I knew what Bubba was like. I knew Bubba doesn't like to admit he's lost, and that if he found one of these under his Christmas tree, he'd be a happy man—if they were $300, of course."

Tuck was drafted by the Army during the Korean War, and received training in "electronic warfare." By 1986, after many years in the tech industry, he had formed a venture capital fund with his business partner, Jim Whitley. They decided their first project would be to form a GPS company. Finding other investors was difficult. Tuck had a simpler vision of GPS than Charlie Trimble's, but it was, in its own way, just as far-reaching in its understanding of the technology. Trimble thought of GPS as a utility to support applications that demanded geographic or temporal precision. Tuck thought of GPS as a personal accessory that spoke to our deepest natures as human beings. "We want to know 'where someone is coming from' and 'where he's going,'" he wrote in Magellan's prospectus. "In the abstract, we speak of a person's or a product's 'position.' We 'pinpoint' things in space and time. The urgency of knowing one's position is even greater than the urgency of knowing the date and time. A person waking from a coma first says, 'Where am I?' then 'What day is it?'" Magellan's products would make locating one's position as easy as glancing at a wristwatch to find the time. "This is not a dumb idea," he insisted.

Tuck was turned down eighty-six times by potential investors, until he found one who agreed to put $500,000 in the venture. He recruited two engineers: Norm Hunt for the software and Don Rea for the hardware. "Little we had done had any relevance to the hardware and software that would ultimately be needed for a low-cost receiver," Rea wrote. "Of course, nobody else had ever done this before either, and we were all too naive to realize it couldn't be done." When Hunt died suddenly during the process, they hired an engineer barely out of college, a software whiz named Valerie Oetting. To complete the image of Magellan as an adventurous company, they recruited Jim Whittaker, the first American to climb Mount Everest, to be the company's CEO.

By 1988, they had a receiver ready, aimed at that old-faithful navigation market, people with boats. Tuck noticed they were developing a very active customer base off the southern coast of Florida—presumably the kind of people who valued the ability to self-navigate on the ocean without getting into a jam that might bring in the Coast Guard. "I'll leave it to your imagination to figure out who," Tuck says.

Trimble was trying to wring every last bit of precision from the GPS signal, marketing products to people like surveyors. Magellan could market an inexpensive receiver because Bubba wasn't a surveyor, or an oil rig captain, or someone looking for precise timekeeping. Bubba just wanted to know where he was, without having to ask another Bubba for directions. "We weren't trying to find your car in the parking lot," Tuck explained. "We were just trying to find the parking lot."

After a few years, Magellan was doing decent business. Then Desert Storm happened. "I don't think it's nice to say that wars are lucky," he said. "But it was lucky that we had one at that particular time—if we were going to have one."

In the years between *Challenger* and Desert Storm, the folly of trying to build a wall around GPS became obvious. Its possibilities sparked the imaginations of too many people for the Department of Defense to maintain the level of control it wanted. GPS was becoming universal.

Nothing was sacrosanct, including the two-channel solution that placed the impenetrable (to most) P code on L2. L2 was not, after all, a private highway—it was more like an HOV/carpool lane that ran parallel to a lane accessible to everyone. L2 was a faster ride, but it was not invisible. It was coming from the same source as L1, making the same curves, traveling in the same direction.

Both GPS codes were coming from the same satellites. The position of the satellites was a matter of public record. L2 was, at root, just another radio signal that could be captured by an antenna, even if its message was gibberish and noise.

Every radio signal has a carrier wave, with the informational component modulated on top of the carrier. Even if you can't understand the information it contains, a close analysis of the carrier wave itself can tell you a lot. It's like receiving a coded letter in the mail and not bothering to crack the code, because the postmark tells you all you need to know. Since the late 1970s, a few scientists had explored using both channels. They discovered that using both resulted in position calculations that were incredibly accurate, sometimes within centimeters. It was a very slow, tedious process, involving a lot of post-processing computer power. By the end of the eighties, people began to think of ways to package this two-channel capability in a consumer receiver. Japan was first out of the gate. "The L2 market," Trimble says, "was opened by the Japanese."

Japan had so far been precocious in adopting GPS for personal navigation, and simple receivers were plentiful in electronics stores. Japan's early lead in developing computerized maps, combined with Tokyo's incomprehensible street address system, made GPS a good fit. The Japanese government now wanted a dual-frequency GPS receiver accurate enough to measure small seismic activity. They approached Trimble and asked if he and his company were up to the challenge. It was a big risk for a small company, as it required considerable time and investment, but Trimble was eager to expand the GPS market. "There were only five satellites in the sky, and not everyone was convinced the system would be completed," he says.

The project also posed a formidable technical challenge. Everyone knew L2 was out there, but picking it up was difficult. "It took us six weeks to find it," Trimble says. "This is how hard the damn thing was." He had nine months to build twenty-five of these receivers, and realized that between the research/development and the manufacturing of these receivers he was risking his small company's future. The contract was very clear that if Trimble was unable to deliver something to exacting specifications, on a hard deadline, the government would walk away and Trimble would have nothing to show for almost a year of work.

Trimble managed to deliver, selling Japan the first dual-channel civilian GPS receiver. Emboldened, Trimble and Eschenbach decided it was time to pursue the American military market. Their competition was the Manpack, the $40,000 cutting-edge military-grade GPS receiver built by Rockwell Collins. The Manpack was portable, but not lightweight—a 25–30-pound cube worn as a backpack, crammed with heavy batteries, 16-bit processors, and clunky hardware.

The Manpack was also weighted down by the sometimes divergent demands of the service branches. The Army complained that when a Manpack was placed in an alkaline bath—a decontamination process in the event of a biological or radiological attack—the alkaline corroded an O-ring, a type of gasket, shorting out the device. Perhaps Rockwell could find a way to omit the O-ring, the Army suggested. That wouldn't do, because the Air Force needed the Manpack to function at an altitude of 50,000 feet, and without the O-ring to release some pressure, the receiver would be an explosive device. "So I said, 'Let's put a pressure relief valve of some kind,' " Gaylord Green recalls. "Oh no, you can't do *that*, because the Navy Seals want to punch it out of their submarine tube. Here was this ten-cent O-ring, and it just sort of epitomized the problems of trying to build the same thing for all three services." (The details mattered; it was a defective O-ring that brought down *Challenger*.)

A small company like Trimble's did not have the kind of clearance required to build military-grade GPS receivers, or access to the process by which public money was funneled to contractors. Trimble had already discovered this when he tried to bid on a project nicknamed "Virginia Slim," the Defense Department's request for a military-grade receiver no larger than a pack of cigarettes. Trimble did not have the resources to "stovepipe" the process of building a receiver, assigning specific components to different companies, a practice which Eschenbach felt compromised efficiency and innovation.

But Charlie Trimble had one thing going for his company, when going up against the Manpack. "I had freedom from a design standpoint that the people at Rockwell and Magnavox never did," he says. "I was designing for what I perceived an end user would want and need—rather than somebody telling me how to approach a problem."

As efficient in its appropriations as America was bureaucratic, Israel was Trimble Navigation's dream client. "It turned out the Israelis have a technique for producing things for the military at a much faster rate than the United States," Trimble says. "They buy something that might work and start testing it. If you're willing to improve it, they buy some more. They got us to a point where we learned how to build a hardened receiver." The final product was a sleek box—an inch high, six inches wide, eight inches deep—far more portable than the unwieldy Manpack. Even better, especially for a soldier in the field, it required very little power to run.

The receiver built for the Israelis became the model for what Trimble called the Trimpack, a.k.a. the SLGR ("slugger"), short for "small light-weight GPS device." Just before retiring from the GPS Joint Program Office, Gaylord Green arranged for the purchase of 200 SLGRs, most of which he planned to divert to the Air Force Academy for training. "I wanted to give Rockwell a little competition, and also energize the commercial GPS business," Green says.

Given its emphasis on ground maneuvers, Trimble felt the Army was the biggest potential customer. The company hired a retired Air Force colonel who had worked with the GPS office to lobby the Army on behalf of the SLGR. The Army asked Trimble how many SLGRs his company could manufacture with a budget of $4 million. Trimble figured they could make a thousand. The Army offered them a contract, impressed by the SLGR's capabilities and by Trimble's streamlined production process. "We could produce the darn things, and the aerospace industry couldn't, because our receivers were simpler than anything the military had," he says.

The civilian GPS industry, represented by Trimble, had successfully barnstormed the military. The Trimpack was portable, durable, and designed with a warrior in mind. It not only provided the latitude and longitude of a user's present location, it could also provide directions to another location, and even store the GPS coordinates of special battle maneuvers.

It couldn't do this very well, though—through no fault of Trimble's. As the new generation of GPS satellites was launched, post-*Challenger*, the Pentagon was finally able to implement its plan to make civilian GPS worse. It was called selective availability, or SA. The operating software for the new satellites intentionally dithered the signal, distorting it just enough so that any position fix was accurate, but not *that* accurate. The margin of error was now about 100 yards. Over time, as new satellites were launched and the first-generation models were removed from the constellation, SA would become universal.

SA was switched on. Almost immediately, the Pentagon had no choice but to turn it off.

Selective availability officially began in April 1990, the same month Trimble delivered the promised thousand Trimpacks. Four months later, Iraq invaded Kuwait—coincidentally, on the same day Trimble Navigation became a publicly traded company—and American troops began pouring into the Gulf. GPS had enough of a profile as something potentially useful to the military that SA became an immediate liability. There were a grand total of thirteen Manpacks available for soldiers in the Gulf. Trimble had just begun fulfilling the Army's initial Trimpack order. The Trimpacks, unlike the Manpacks, could not use the military's code. In August, after just four months, the Pentagon quietly turned off SA, so that the Trimpacks could work to their full potential.

The Army added to its initial Trimpack order—both a boon and a curse for Trimble. "We were a $60 million-a-year company, trying to produce a thousand of these things a month, and we did not have automated production," he says. The price of the inexpensive cellular telephone parts Trimble used in the Trimpack, acquired from Japanese suppliers, skyrocketed during the final quarter of 1990, as the Japanese cellular telephone industry exploded. Trimble was now losing money on every Trimpack the company produced. They didn't even have an ironclad contract with the military—just a verbal request that they keep 'em coming. Charlie Trimble felt a patriotic obligation, so he ratcheted up production, ultimately producing around 4,000 by the time the war began. It was the Trimpacks, most of which ended up in tanks, that allowed Coalition forces to advance confidently across the desert.

The majority of soldiers were not lucky enough to procure a Trimpack. But word was spreading through the forces about GPS. Trimble was operating at capacity—there was no way the company could make more Trimpacks. Magellan—the only other company in the fledgling civilian GPS industry to churn out portable receivers—was only too happy to step in and pick up the demand. Were their receivers as accurate as Trimpacks? Not a chance. But they could offer users a reasonable position fix out in the endless, featureless, directionless desert. (You go to war with the GPS receivers you have, not the ones you wish you had.)

Despite a price tag of $1,000, soldiers wanted those Magellans—even if they had to pay for them out of their own pockets. Their families called the company, which directed them to marine retail outfits. The most enterprising soldiers would manage to call Magellan directly from the Gulf and arrange for the company to ship the receivers directly to them. During the run-up to Desert Storm, there were sightings of C-130 military transport aircraft landing at NASA's Ames Research Center in Silicon Valley; the crews would make a run for a Bay Area electronics store, scouring the shelves in search of Magellans.

After the war ended, the Pentagon was of two minds regarding the civilian GPS industry. While publicly praising it—one longtime Trimble employee remembers people approaching Charlie during a trip to Washington, shaking his hand, and thanking him for his service—tension over what Trimble and others had done began to surface, especially regarding the dual-frequency receivers. "There was a cabal inside the military that was really unhappy with what we were doing with survey-grade GPS," Trimble says.

Javad Ashjaee, Trimble's former colleague, was one of those fanning the flames. In late 1991, his company, Ashtech, advertised a receiver that it claimed was "far more accurate" than the military-grade receivers Rockwell Collins was manufacturing. How had he done this? Ashjaee wouldn't say, but the industry rumor was that he had decoded at least part of the military signal. "We never even partially decoded it," Trimble says of the P code. "We considered that as weakening military equity. And so the military got plenty unhappy with Javad."

"Ashtech did some things that may have been borderline wrong," said Scott Pace, the professor at George Washington University who, in the early 1990s, monitored the GPS industry from his post at the U.S. Department of Commerce. "There was an allegation that they figured out how to access the encrypted code." Jules McNeff, the officer who defended the Air Force budget at the Pentagon in the late 1980s, will only say that Ashjaee "exploited some knowledge about the signal structure that he probably shouldn't have."

"This is the first time I'm hearing about it," Ashjaee says two decades later, while referring to McNeff as "my good buddy," apparently sincerely. "We had no connection to the military. We did not use military information."

The Defense Department decided that because GPS was foremost a military technology, anyone who marketed GPS products was a munitions maker. This made commercial GPS receivers subject to export controls that effectively closed off huge international markets. "The Pentagon apparently intends to 'reward' the companies that rushed to fill its emergency orders by forcing them to concede an estimated $3 billion worldwide market to European and Japanese competitors," the *Bulletin of the Atomic Scientists* grumbled.

"It wasn't that anyone was being hostile—the government was just ignorant," Pace says. "They were thinking that L1 code is civil, and the L2 is 'military.' One fish, two fish, red fish, blue fish—a Dr. Seuss level of understanding. But that can't be right, because all you're doing is ionospheric correction with the second signal. Is that inherently a military activity?"

Pace urged the major players in the industry to speak with one voice. "At the time, the GPS industry was largely made up of tiny companies that were fiercely competing with each other," Trimble says. "The blood between ourselves and Ashtech was certainly on the bad side of neutral. And Magellan . . ." He trails off. Tuck says Magellan's attempt at an IPO in 1991 failed "partly because Trimble had badmouthed us all over the place to the investor community." The major players put aside their differences and met with the Pentagon to emphasize their commitment to helping the military protect its stake in GPS. Their concessions included software tweaks that would prevent their receivers from functioning if they were moving faster than 1,000 miles per hour or at an altitude above 60,000 feet, lessening the likelihood of someone using a GPS chip to build a homemade smart weapon. Nobody had any objection to the ongoing encryption of the military's GPS signal, a practice denoted by the acronym AS, short for anti-spoofing. (Well, almost nobody—Ashtech's "Kiss Your AS Goodbye" corporate ads suggested a responsible opposing viewpoint.)

Left unsettled was the status of selective availability. The Pentagon kept the policy in place while quietly discontinuing it during times of U.S. military activity, so that soldiers could continue to use civilian-grade receivers. When American troops intervened to remove Haiti's military government in 1994, attentive GPS users worldwide noticed that their receivers gave more accurate readings.

What ultimately doomed selective availability was the existence of a supplementary positioning technique called DGPS—the D is for "differential." DGPS is not a specific technology, but rather a general concept that involves comparing a fixed location's known coordinates to the same location's coordinates as computed by a GPS receiver. The discrepancy reveals the GPS error at that location, allowing for a continuous refinement of the readings from nearby GPS receivers. In a DGPS network, one or more continuously operating GPS receivers broadcasts the correction to any GPS receivers in the area authorized to receive the correction. These receivers adjust their calculations accordingly.

Even absent the intentional errors of selective availability, DGPS is a useful way to strengthen the accuracy of GPS readings. Nations began to set up their own DGPS networks, particularly around coastal areas, where they served to make GPS more accurate for mariners and improve international trade. In the United States, this created the odd situation of one branch of the Armed Forces strengthening GPS for civilians, via a DGPS network administered by the Coast Guard, while the official military policy was to degrade civilian GPS through selective availability.

With each passing year in the 1990s, the continued development of differential GPS methods made selective availability look increasingly worthless—a point Javad Ashjaee made by running an Ashtech ad featuring the *Mona Lisa* with various parts missing—the idea being, why would you willfully corrupt the masterpiece that is GPS? "Back in those days, it had some legitimacy in terms of the uncertainty it afforded to adversaries," McNeff says. "But at some point, you have to say we're fighting yesterday's war, we need to look at how we live in this new environment and deal with it. Leaving it on was more of a hindrance—and, frankly, an impediment to having the military move out and develop things they needed to do to live in that environment." In 1995, a report from the National Research Council strongly recommended eliminating SA because "any enemy of the United States sophisticated enough to operate GPS-guided weapons will be sophisticated enough to acquire and operate differential systems."

It was ultimately the lure of an untapped market that finally banished selective availability. Brad Parkinson publicly commented that not only was the policy ineffective, it was inhibiting the development of commercial GPS applications: personal navigation, for example (there was a growing interest in putting GPS guidance systems in cars). Garmin, founded in 1989, would soon build on Magellan's idea of personal navigation for the Bubbas of the world, and make that market explode. Turn-by-turn directions required more precision—football field-length miscalculations would not do.

By the middle of the decade, the market for military GPS was flat compared to the rapid growth of the commercial GPS industry—worth $2 billion in 1996, and growing at an annual rate of 20 percent. That year, President Bill Clinton, in consultation with the Department of Defense, announced in a presidential decision that the selective availability policy would end within ten years. At midnight Eastern Daylight Time on May 1, 2000, the intentional degrading of the civilian signal ceased. "This will mean that civilian users of GPS will be able to pinpoint locations up to ten times more accurately than they do now," Clinton said in a statement. The death of selective availability was catalytic for the GPS industry. Within two weeks, Garmin's sales were up 40 percent and Magellan's retailers were setting company records.

Gauging the overall value of GPS is nearly impossible. In 2011, when members of the GPS regulatory and scientific community mobilized against plans to authorize a private wireless network they feared would threaten the GPS signal, several cited the barely fathomable figure of $3 trillion as the market's value. It has become difficult to untangle the worth of GPS from the worth of *everything*. In an increasingly cloud-based world, the global market for the so-called "Internet of things"—the ability for physical objects (including people) to exchange data over cloud-based networks—could reach $1.5 trillion by 2020. These systems often require location information, which will be provided by GPS—and time synchronization that will also likely be tied to GPS. Placing an economic value on GPS has become nearly as impossible as pegging the value of other utilities. How much money do electricity and telephones generate? How much is oxygen worth to the human respiratory system?

As receivers have come to integrate other calculations to augment GPS—a continued goal that is actually codified in America's National Space Policy—plain old civilian GPS just gets stronger. This is something that Javad Ashjaee saw very early, when he decided to create receivers that used both GPS and GLONASS, a risky venture given the political and economic upheaval Russia has experienced. When Ashjaee's partners worried that a continued GPS/GLONASS collaboration would negatively affect attempts to take Ashtech public, he left his own company to found another, Javad Positioning. In the U.S., Ashjaee remembered, "supporting GLONASS was an unpatriotic act. The most prominent figures of GPS teased me for wasting my time with GLONASS."

GLONASS is nowhere near as dependable as GPS. But its extra satellites add a boost to GPS readings. Ashjaee isn't exaggerating when he claims that the fastest real-time ultraprecise GPS position calculation "is not possible without combining GPS and GLONASS satellites." It's certainly true for beets.

"On all our stuff, we've got GLONASS unlocked," Troy Seaworth said. We were standing in the high-ceilinged garage at Seaworth Farms, just outside Fort Collins. He was explaining how many passive ranging signals are necessary to keep an automated tractor following a path with enough precision for beet farming. The four simultaneous signals GPS guarantees at any given time are not nearly sufficient, so Seaworth Farms' equipment combines GPS readings with other systems, including GLONASS. "Usually, we're right around twelve satellites," he explained. "If you get below six or eight, you can tell."

Seaworth Farms has been in the family since Troy's grandfather purchased the 2,000 acres in 1945, using most of it to grow beets, which are still the farm's main crop, along with corn, hay, and beans. In the mid-1990s, when Troy was studying agronomy at Kansas State University, he learned about the burgeoning field of precision agriculture. Practitioners of precision agriculture squeeze every last bit of efficiency out of the process, maximizing yield while minimizing cost—in money and time—and the use of resources such as water, fertilizer, and pesticides.

Troy became a convert, and convinced his dad that investing in precision agriculture would quickly pay for itself. Troy wanted to convert most of the farm to a farming practice called strip till. The process of harvesting a field leaves behind leaves and other organic residue, which a farmer would traditionally remove in preparation for the next planting. With strip till, the detritus remains. The field is tilled, with each row separated by a small amount of the fallow area. Those rows of untilled land hold down the soil, protecting it and keeping it cool so that it retains moisture. A strip-tilled field can hold fewer crops, but the overall yield will actually be greater than if the entire field is planted.

The catch is that all rows must be bounded by *exactly* the same amount of fallow area. Otherwise, a tractor dragging a planter or fertilizer dispenser that handles several rows simultaneously will miss the rows entirely. Accordingly, that means that the planters and fertilizer dispensers must also be exactly the same width. And the tractor driver had better drive in exactly a straight line, a challenge even for experienced farmers, especially on fields with no landmarks with which to orient yourself. One of the earliest precision agriculture products was a strip of lights placed on the dashboard. A GPS receiver would determine the headings. The driver aimed to keep the strip lit in the center; any drift, and the lights on that side would illuminate.

It was spring, the day Troy had been planning to plant, but a freak blizzard the night before delayed him by a day. Troy showed me around the garage, which housed two tractors and the skeletal frames of planters, which resembled giant combs with lots of space between the teeth, outfitted like cyborgs with wires and cables that connect to the tractor's rooftop receiver and the map data it contains. More wires protruded from the tractors' cabs, which had computer monitors next to the driver's seat.

Sporting a goatee and mirrored shades hanging on his collar, revealing the permanent squint of someone who spends a lot of time outdoors, Troy spoke of precision agriculture with a computer geek's enthusiasm. Map software allows him to analyze crop yields around the farm, overlaid with a soil analysis map, and then program the tractor to drop fewer seeds on the less fertile patches. Later, the system will drop less fertilizer on those areas. "That thing knows exactly how fast you're going, and it knows exactly how much you need to put down," Troy said. "We can spray 500 acres a day, and we'll run out right at the end of the field. It's pretty unbelievable."

The planting process is almost fully automated. The tractor drives itself, hewing to its prescribed GPS coordinates. All the driver has to do is turn it around at the end of a row, and it snaps back into position. Troy often uses the time to check the commodities markets on his phone.

Precision agriculture is especially helpful in growing the notoriously finicky sugar beet, especially at harvest time, when the system knows the location of every beet. "When you're digging, if you move just this far off," he said, moving his thumb and forefinger an inch apart, "you're slicing the beet and breaking it. The way Dad used to do it, you're driving the tractor, and with hydraulics you could move the beet digger a little bit. With one guy, it's pretty hard to do. So back in the sixties, they would have a seat on the back, so one guy could steer the digger while the other guy drove the tractor. Whereas now, GPS drives the tractor, and I don't get off the row because we have sub-inch accuracy. We need to be that close, because once you cut 'em you can't dig 'em, because they just break. They're gone."

It is in pursuit of this sub-inch accuracy that Seaworth Farms' precision agriculture system includes dual-channel GPS receivers that receive constant real-time broadcasted correction from a stationary receiver nearby. This is the legacy of the experiments with the military code Trimble and other conducted. What once required post-processing can now be computed and broadcast as it happens. The tractors are constantly receiving small correctives, tiny adjustments to their directional heading, so that they delicately lift beets from the soil.

Between 2006 and 2012, usage of GPS-based precision agriculture tactics worldwide tripled; the 2.5 million farms in the U.S. and Canada accounted for more than half of overall usage. In the U.S. alone, during that period, GPS was probably worth around $20 billion to farmers, in both decreased input costs (soil, fertilizer, etc.) and increased output—a figure that has probably risen to about $33 billion today.

The adoption rate in Europe, while growing, has lagged behind the rest of the world. Just 22 percent of farms in the U.K. are using navigation satellites to increase accuracy, though that figure is slowly increasing. A survey of precision agriculture practices across the continent commissioned by the European Parliament noted a widespread skepticism of these methods among farmers in England, concluding that "there is a long way to go before the majority is convinced."

By 2020, the precision agriculture market (including tools that do not use GPS) will be worth $4.5 billion, and nearly 50 percent of the world's tractors will use GPS. The fastest-growing market for GPS-based precision agriculture is Asia and the Pacific, where the adoption rate has been almost viral, from nearly nobody in 2006 to 17 percent of the world's installed GPS precision agriculture systems in 2015, and growing at nearly twice the global rate. Evidence suggests that precision agriculture may be just as helpful for tiny farms (one or two acres) in developing nations as it is for larger farms. Small farmers can use simpler tools, such as handheld soil analyzers, rather than the expensive multi-satellite real-time corrective systems like Seaworth's.

Even on the smallest level, GPS is becoming part of a precision agriculture approach. In Uttar Pradesh, India, farmers level their fields by dragging a wooden plank behind an ox. The result is uneven, leading to an uneven distribution of water. An experiment conducted on a two-acre farm leveled half the land with the traditional method and half with a small precision leveler using GPS. The precisely leveled side yielded 2.25 metric tons of spring wheat, an improvement of nearly 300 percent over the ox-leveled side.

As complex as a system like Seaworth's is, it paradoxically lessens the gap that separates a farm like his from a two-acre field in Uttar Pradesh. A detailed soil variability map is, in a sense, the rebirth of a cognitive map that came with farming. "The original way, what my grandfather thought was farming, was he would walk behind a horse," Troy said. "He was farming a very small area of land, but he knew what the variability was, because he was down there looking right at it. As we've become more mechanized, it's taken away that site-specific knowledge we used to have from walking behind a horse."

Troy pointed at the mutant tractor, and the amalgamation of wires, sensors, and controls that formed the terrestrial end of an obsessively precise space-based positioning system. "This is giving us back the intimate knowledge we had when we were much smaller and more intimate with the land," he said. "I think we're coming full-circle to the knowledge we used to have."

Captain Cook launched his third and final Pacific voyage in July 1776, navigating the ocean with the help of his chronometer. He still had not found an answer to the question he had posed in Tahiti, six years earlier, regarding Polynesian migration—"How shall we account for this nation spreading itself so far over the ocean?"—or confirmation of his theory, hatched on Ra'iatea, that the migration had proceeded from the west. "When this comes to be prov'd," he'd written, "we shall no longer be at a loss to know how the Islands lying in those Seas came to be people'd."

On Atiu, an island in the chain that would later be named the Cook Islands, he heard a story about how some of the island's inhabitants had left Tahiti, bound for Ra'iatea, and accidentally drifted to Atiu. So maybe that was the story all along—the Polynesians had settled these islands at the whims of the currents and winds. "This circumstance very well accounts for the manner the inhabited islands in this Sea have been at first peopled; especially those which lay remote from any Continent and from each other," he wrote.

After more than a year at sea, Cook became the first documented European to arrive at an island chain in the far northern end of the Polynesian triangle, one of the most remote inhabited places in the world, which he decided to name the Sandwich Islands. His ship, the *Resolution*, landed at Kealakekua Bay on the island of Hawai'i. The *Resolution* left a month later, but soon turned back to repair some damage.

What happened next is still not fully understood. A longstanding theory is that because Cook's return coincided with a harvest festival dedicated to the god Lono, the island's inhabitants may have ascribed godlike powers to Cook, and his return, after the season of Lono had passed, caused a reappraisal of his deification. Whatever the cause, tensions arose between Cook's crew and a crowd they encountered on the beach. In the ensuing hostilities, Cook was stabbed several times. He died facedown in the surf. At least one account claimed that his chronometer stopped ticking at the precise moment he perished.

During the *Resolution*'s first visit, some local chiefs had come aboard to meet Cook. One of them, barely twenty years old, may have even stayed overnight on the ship at the invitation of the crew. In the years following Cook's death, this chief, Kamehameha, would lead a brilliant and bloody campaign to unite the islands in the archipelago as the Kingdom of Hawai'i, the final independent Polynesian nation-state in this farthest-flung part of the region, one of the last destinations reached by the Polynesian migration, and therefore possibly the last part of the world settled by humans. The kingdom lasted until it was overthrown by non-Hawai'ians in the late nineteenth century, and was later annexed to the United States.

The island that Cook visited is today commonly known as the Big Island. On the other side of the island from where Cook met his death, there is a trail, difficult to locate but not that hard to reach, that winds along the coast, on cliffs above the ocean. The trail, wide enough for off-road vehicles, has eroded into kind of a wash, sunk a few feet below the brush and lava rocks on both sides.

The trail eventually leads to a *heiau*, a Hawaiian temple and sacred space. It is mostly a wall of lava rocks surrounding an area of about an acre, with more rocks arranged in various patterns inside. Kamehameha had this built as a tribute to the war god Ku, acting on a prophecy that its construction would lead to him unifying the island. To move the rocks to the site, men loyal to him formed a line that stretched for several miles. It was here that Kamehameha killed his cousin, a leader of rival forces. Today, the *heiau* remains preserved; the walls of rocks have not crumbled.

Walk further down the path, and there is a similar site, this one much smaller. It marks the place where Kamehameha may have been born.

The path ends soon after this, running up against a fence that marks the edge of a private ranch. But not far away, maybe a half mile across fields, on bluffs with a pristine view of the ocean, there is a radio tower. It is one of just eighty-five, placed along United States coastlands, that form the U.S. Coast Guard's Nationwide Differential GPS Service. Satellite images reveal a cleared area around the antenna, not unlike the one around the *heiau*. From the air, it looks like a shrine. Which, in a way, it is.

© 2015 Google, © 2015 TerraMetrics

You Have Arrived

CHAPTER FIVE

Death by GPS

One early morning in March 2011, Albert Chretien and his wife, Rita, loaded their Chevrolet Astro van and drove away from their home in Penticton, British Columbia. Their destination was Las Vegas, where Albert planned to attend a trade show. They crossed the border and, somewhere in northern Oregon, they picked up Interstate 84.

The straightest route would be to take I-84 to Twin Falls, Idaho, near the Nevada border, and then follow U.S. Route 93 all the way to Vegas. Although U.S. 93 would take them through Jackpot, Nevada, the town near the Idaho state line where they planned to spend the first night, they looked at a roadmap and decided to exit I-84 before that junction. They would choose a scenic road less traveled, Idaho State Highway 51, which heads due south away from the I-84 corridor, crossing the border several miles to the west. The Chretiens figured there had to be a turnoff from Idaho 51 that would lead them east to U.S. 93.

Albert and Rita had known each other since high school. During their thirty-eight years of marriage, they had rarely been apart. They even worked together, managing their own small excavation business. A few days before the trip, Albert had purchased a Magellan GPS unit for the van. They had not yet asked it for directions, but their plan wasn't panning out. As the day went on and the shadows grew longer, they were not finding an eastward passage. They decided it was time to consult the Magellan. Checking their roadmap, they determined the nearest town was Mountain City, Nevada, so they entered it as the destination into their GPS unit. The directions led them onto a small dirt road near an Idaho ghost town and eventually to a confusing three-way crossroads. They chose the one that seemed to point in the direction they wanted to go. And here their troubles began.

If Albert had been navigating the route in the daytime, he might have noticed that it was taking them through the high desert as it rose toward shimmering snowy peaks in the distance. In the dark, the changes were so subtle that they barely registered. And besides, he *was* on a road—"a pretty good road," the Elko County sheriff would later say, that "slowly goes bad." Through the night, it carried them higher into the Jarbridge Mountains, deeper into the backcountry. The road twisted, dipped, rose again, skirting canyons walled with sagebrush. It was the kind of terrain for which the Chretiens' van was not designed.

Several days passed before their family and friends realized that Albert and Rita had never arrived at the trade show. The couple had not informed anyone of their detour, so nobody knew where to look for them. The manhunt involved police agencies in four states, scouring 3,000 miles of highway, with the most intense efforts in eastern Oregon, where they had used a credit card in a convenience store. On April 8, just shy of three weeks since Albert Chretien left Highway 51, authorities announced they were scaling back search and rescue efforts, a tacit admission that wherever the Chretiens had gone, it was too late to find them.

What happened to the Chretiens is so common in some places that it has a name. The park rangers at Death Valley National Park in California call it "death by GPS." It describes what happens when your GPS fails you, not by being wrong, exactly, but often by being *too right*. It does such a good job of computing the most direct route from Point A to Point B that it takes you down roads which barely exist, or were used at one time and abandoned, or are not suitable for your car, or which require all kinds of local knowledge that would make you aware that making that turn is bad news.

Death Valley's vast arid landscape and temperature extremes make it a particularly dangerous place to rely on GPS. In the summer of 2009, Alicia Sanchez, a twenty-eight-year-old nurse, was driving through the park with her six-year-old son, Carlos, when her GPS directed her onto a vaguely defined road that she followed for 20 miles, unaware that it had no outlet. A week later, a ranger discovered Sanchez's Jeep, buried in sand up to its axles, with sos spelled out in medical tape on the windshield. "She came running toward me and collapsed in my arms," the ranger wrote in a report. "Her lips were very dry and chapped with bleeding blisters and her tongue appeared to be swollen with very little saliva formation. I walked over to the Jeep and looked inside. I saw a boy slumped in the front seat with obvious signs of death." Mother and son had wandered over ten miles of desert in search of water, and had resorted to drinking their urine. They had tried to share a Pop-Tart a few days earlier, but their mouths were too dry to swallow. As he lay dying, Carlos grew delirious, telling his mother he was "speaking to my grandfather in heaven."

Most death-by-GPS incidents do not involve actual deaths—or even serious injuries. They are accidents or accidental journeys brought about by an uncritical acceptance of turn-by-turn commands: the Japanese tourists in Australia who drove their car into the ocean while attempting to reach North Stradbroke Island from the mainland; the man who drove his BMW down a narrow path in a village in Yorkshire, England, and nearly over a cliff; the woman in Bellevue, Washington, who drove her car into a lake that their GPS said was a road; the Swedish couple who asked GPS to guide them to the Mediterranean island of

Capri, but instead arrived at the Italian industrial town of Carpi; the elderly woman in Belgium who tried to use GPS to guide her to her home, 90 miles away, but instead drove hundreds of miles to Zagreb, only realizing her mistake when she noticed the street signs were in Croatian.

These types of mishaps often elicit sheer bafflement. The local Italian tourist official noted that although "Capri is an island," the unfortunate Swedes "did not even wonder why they didn't cross any bridge or take any boat"; the first responders in Bellevue were amazed that the women "wouldn't question driving into a puddle that doesn't seem to end." For their part, the victims often couch their experiences in language that attributes to GPS a peculiar sort of agency. GPS "told us we could drive down there," one of the Japanese tourists explained. "It kept saying it would navigate us a road." The BMW driver echoed these words, almost verbatim: "It kept insisting the path was a road."

Something is happening to us. Anyone who has driven a car through an unfamiliar place can attest to how easy it is to let GPS do all the work. That GPS can have a transformative effect on a society is undeniable. Claudio Aporta, an anthropologist, has studied the use of GPS among the Inuit who live in the Igloolik region of the Canadian Arctic. Over many generations, the Inuit have developed complex wayfinding techniques to maneuver through a vast, often frigid landscape. The first GPS units began to appear there in the late 1990s, and were mostly used by hunters. Today, nearly every family has at least one GPS device.

For some in the community, especially the old residents, GPS, much like snowmobiles and computers before it, has made life more convenient at the cost of cutting off its users from revered traditions and the land itself. For others, GPS is not a break with tradition, but rather its extension, because it maximizes one's ability to move through the world. Aporta observed that using GPS can lead to a sense of "disengagement," because the question of location, which once required a close interaction with the world, is now solved by unseen technologies far removed from the user. But total disengagement is never an option. "There is important environmental information that people still need to travel, concerning not only the state of the sea ice, but also the conditions of snow and wind," he says.

In our society, total disengagement is often an option. We have come to depend on GPS, a technology that, in theory, makes it impossible to get lost. Not only are we still getting lost, we may actually be losing a part of ourselves.

In 1948, nearing the end of his career, Edward Tolman, a UC Berkeley psychology professor, published an article called "Cognitive Maps in Rats and Men" in the journal *Psychological Review.* He summarized several recent experiments, conducted in his department and elsewhere, that studied the wayfinding behavior of rats wandering mazes in search of a reward. The findings, Tolman argued, had much to teach us about how humans navigate and orient themselves.

Tolman believed there were two main schools of thought among his colleagues. One group believed that a rat's central nervous system "may be likened to a complicated telephone switchboard" that helps the rat compile stimulus–response connections in its brain. When a rat is confronted with a new maze, the switches are closed. As the rat explores the maze—making decisions, reaching dead ends, doubling back for another approach—the switchboard lights up. It receives incoming calls from the rat's sense organs and relays the message to the animal's muscles. The moves that a rat discovers take it closer to the reward are the switchboard connections that remain open, while the incorrect turns are the switchboard connections that close.

Tolman was openly contemptuous of this stimulus-and-response school of thought, which he suggested imagined a "satisfaction-receiving part of the rat" that tells the switchboard operator to "hold that connection; it was good; and see to it that you blankety-blank well use it again the next time those stimuli come in." Tolman counted himself among the second group, the field theorists. For them, the rat's nervous system was not a switchboard, staffed by a neutral operator, but rather a "map control room," the domain of an active cartographer using sense-organ stimuli to construct a "tentative, cognitive-like map of the environment." Learning was not merely compiling a list of which moves served the rat's purpose and which did not. It was an ongoing process of adding detail to the cognitive map, increasing the animal's ability to perceive the maze as a totality.

Tolman divided the cognitive map concept into two basic models, strip maps and comprehensive maps. A strip map is analogous to a visual map that depicts only the spatial relationship between two points: an unbroken line surrounded by blank space on the paper. As the cognitive map gains depth, breadth, and context, it becomes a comprehensive map, so that the organism can now visualize the orientation of point A to point B, point B to point C, point A to point C, and so on. The more intimately one knows an environment, the more details the cognitive cartographer can add to the map. Tolman argued that the rat experiments suggested that the animal had the capacity to build strip maps into comprehensive maps.

As the first major neobehaviorist, Tolman sought a rethinking—but not a rejection—of behaviorism's core tenets. Behaviorism had swept the discipline in the early part of the century, a reaction against psychology's emphasis on abstract concepts of mind—"making *behavior*, not *consciousness*, the objective point of our attack," John Watson proclaimed in 1913. The behaviorists were influenced by a prior generation of psychologists, such as William James and John Dewey, for whom Darwin's evolutionary theory was a touchstone. Thoughts existed to motivate some actions and discourage others, but it was the actions that mattered.

Tolman argued for a behaviorism that considered cognition as more than just synaptic connections, one that considered abstract concepts (consciousness, cognition) as well as observables (action, behaviors), by designing experiments that allowed organisms to demonstrate, via their actions, the structure of their thinking. Instead, behaviorism became more doctrinaire. When Tolman published "Cognitive Maps in Rats and Men," the most influential behavioral psychologist was Clark Hull at Yale, who argued that every action an organism performs, down to the level of thought itself, is governed by stimulus and response. At Harvard, B. F. Skinner succeeded Hull as behaviorism's leading thinker. One of the most attractive aspects of Skinner's work was its empiricism—its precepts could be explored in controlled conditions, such as the "Skinner box," which tested the extent to which one could condition an animal's behavior—and its implication that total control over behavior was possible.

Tolman was wary of theoretical certainty, and was under no illusion that these rat experiments proved the existence of the cognitive map, which was, after all, just a metaphorical model for understanding behavior. But it was a model, he felt, that epitomized conscious experience. Tolman projected himself into the maze, empathizing with the rats on two levels, as both a sentient being and a scientist. What is the world if not a maze through which we all navigate, using the tools and maps—cognitive and otherwise—we have at our disposal? And what are scientists if not rats in a maze of inquiry, assembling knowledge and testing it against the observable environment? Science produces "a map and a picture of reality," nothing more or less. "If it were to present reality in its whole concreteness," Tolman wrote in 1932, "science would not be a map but a complete replica of reality. And then it would lose its usefulness." Perhaps Tolman was influenced by Alfred Korzybski's dictum, issued a year earlier: "The map is not the territory." Even the most comprehensive cognitive map is not the world, which is always mediated by our perceptions of it. There is no escaping the maze.

Beginning in the early 1970s, Tolman's cognitive map concept experienced a renaissance, adopted by experimental and developmental psychologists, geographers, urban planners, architects, and anyone interested in wayfinding. (Thomas Gladwin used it when discussing Carolinian navigators' *etak* system.) Interpretations of the term vary, but most treat the cognitive map as something more than symbolic but less than empirically observable, as in geographer Rob Kitchin's definition: "a map-like representation within the 'black box' of the nervous system." Many also recognize two or more typologies that echo Tolman's strip maps and comprehensive maps.

In a fascinating article that appeared in the Canadian journal *The Walrus* in 2005, journalist Alex Hutchinson described recent developments in cognitive science that are particularly relevant in the age of GPS. Among them was the definition of a rare neurological disorder called developmental topographical disorientation (DTD), which seems to prevent sufferers from forming even simple cognitive maps, so that they require years to master routes as repetitive as daily commutes. One DTD researcher interviewed by Hutchinson argues that all wayfinding involves some combination of cognitive mapping and stimulus response—even if we visualize the terrain, we may also rely on memorized cues while traveling—and that those with the condition have recourse only to the latter.

If neurological pathologies can restrict cognitive maps, what about environment? After arguing for the existence of the cognitive map as a wayfinding tool, Tolman expanded the concept such that the *idea* of the cognitive map was itself a worldview. His article ends with a "brief, cavalier, and dogmatic" suggestion: perhaps some of the experimental setups that limit a rat's ability to progress from strip maps to comprehensive maps, such as overly frustrating conditions, correlate with antisocial behaviors in humans that "can be interpreted as narrowings of our cognitive maps." Displaced aggression, for example—the projection of personal frustrations onto outsiders misperceived as the cause of one's problems, such as "poor Southern whites, who take it out on the Negroes," "[Berkeley] psychologists who criticize all other departments," even "we Americans who criticize the Russians and the Russians who criticize us"—Tolman saw as an example of a stunted cognitive map, a blindered, strip-mapped conception of the world. "My only answer," he wrote, "is to preach again the virtues of reason . . . broad cognitive maps," and pray that conditions favorable to their development would prevail across "the great God-given maze which is our human world."

Which brings us to GPS, a technology Tolman never lived to see. What might he have thought of our devotion to GPS as a means of making our way through this maze? The soothing voice of the turn-by-turn directions, guiding us through an unfamiliar environment, is the personification of the strip map. Allegiance to that strip map promotes the reasoning that lies behind the most baffling death-by-GPS scenarios, the willingness to "turn right here" when "here" is clearly a lake. But the specter of the strip map also haunts the scarier incidents, like the disappearance of the Chretiens—scary because they cause actual death, but also because we can imagine them happening to us, like a random car accident. Who can fault Albert Chretien for following a road, even if he lacked a map comprehensive enough to know it was a road best not taken? Let those who have never experienced the liberating sensation of switching on GPS and switching off their navigational awareness cast the first stone.

The rise of personal GPS devices—including the integration of GPS with mobile phones—has been so meteoric that we are just now beginning to take stock of how GPS can affect the cognitive map. We may be witnessing the mass narrowing of the human cognitive map—as a construct (a decrease in navigational ability), but possibly also on a more literal level, an actual reordering of our neurons. If we are indeed shrinking our cognitive maps, we are doing it to ourselves, our love of GPS creating the conditions that lead to their narrowing.

This is something Tolman did not foresee in his plea for global conditions that promote "reason . . . broad cognitive maps": a world actively creating a condition that discourages them. And the strangest thing is, this desire to enhance the experiencing of driving by increasing the driver's alienation from the environment seems to predate the automobile, suggesting that this desire, whatever its roots, runs deep.

Taking a long enough view, and a broad enough definition of the concept, the first "GPS" auto navigation unit was the "south-pointing carriage," which appeared in China around 2,000 years ago. Whenever the carriage changed direction, a gear-driven mechanism would measure the movement of the wheels and cause a figure with an outstretched arm to always point in the direction of the carriage's original heading. In the early twentieth century, users of the Jones Live Map for cars could "program" routes on a paper disc mounted on a turntable, which was connected to a gear train attached to one of the car's wheels. As the car moved, the disc spun, revealing printed directions that told the driver when to turn. A competing technology, the Chadwick Road Guide, used a route disc linked to the odometer, which triggered color-coded signal arms that alerted the driver to an upcoming maneuver.

In the 1960s, engineers began to experiment with large-scale car navigation systems, using proximity beacons spread around an area, transmitting location-coded signals. The first such system, which employed roadbed magnets and binary code to communicate location information to passing vehicles, was developed and tested by General Motors, though never implemented. The systems that did appear were mostly implemented by governments, which saw them more as ways to mitigate traffic than as personal navigators. When using the U.S. Federal Highway Administration's Electronic Route Guidance System (ERGS), developed in the late 1960s, drivers entered destination codes into an onboard console. As the cars approached certain intersections, the code was transmitted to roadside units that analyzed routing options and transmitted a suggested route back to the console. The Japanese CACS system, launched in 1973, used 100 beacons placed at major intersections in southwest Tokyo, linked to a central computer. When a CACS car passed one of the intersections, it would transmit its route and receive guidance information and an estimated travel time to the next link. In the late 1970s, West Germany briefly experimented with a similar system.

By the 1980s, proximity beacon experiments were mostly abandoned in favor of dead reckoning systems that detected location using odometer information and other data. Nolan Bushnell spearheaded the most ambitious attempt. Bushnell was already a tech success story, someone who had changed the firmament of the amusement industries by creatively leveraging technology. Atari, the company he cofounded in the early seventies, marketed simple video games that utilized chips and logic gates, but no microprocessor—essentially, computer games that required no expensive computer—and had a huge hit with Pong. These games first appeared in bars and taverns—as with games like darts and pool, they were an inducement for customers to stay and order more drinks—but their popularity soon merited their own dedicated arcade spaces. Bushnell's next step was to vertically integrate by moving these games back into taverns. Except now, the "taverns" would be his own chain of pizza parlors, Chuck E. Cheese's Pizza Time Theatre, heavily stocked with video games, their incessant bleeps and chirps competing with the sound of "live" music performed by a band of anthropomorphic animatronics.

By 1983, he was ready for another challenge. While participating in a sailing race from California to Hawai'i, he shared a boat with Stan Honey, a master sailor and navigation enthusiast who did research on military systems, such as sensors and radar, for SRI International, a Stanford-affiliated think tank. Out on the ocean late one night, the two got to talking about navigation systems—specifically, what it would take to build one for cars. Unlike boats, cars travel on a network of interconnected lines called streets, so there was an inherent manageability. A dead reckoning mapping system should be fairly easy to design, they decided. Honey, an admirer of traditional Pacific wayfinding, suggested they call the company Etak.

Etak's innovation was to augment dead reckoning with map-matching algorithms that allowed the system to compare physical locations with digital map data. A car outfitted with an Etak system had special tire rims that provided a more accurate read than a standard odometer, and the distance traveled was calculated based on wheel rotations. A compass kept track of the car's direction. For the map-matching component, Etak took publicly available maps compiled by the U.S. Census Bureau, which contained street address information. Since these maps were extremely imprecise, Etak used powerful Vax computers to match the streets in the census database with aerial photographs. These corrected maps were stored on audiocassettes for use in the car. A five-inch screen mounted on a stalk inside the car displayed a simple moving map—glowing green lines against a black backdrop—that displayed the car's location.

Etak was a hit in Germany and Japan, but American car companies were a tougher sell. "Detroit was so tone-deaf," Bushnell says. "The car companies were all brain-dead. We could show them that they could sell navigation systems all day long for $1,500 to $2,000, with an installation time of just two hours. I have no idea what their little minds were thinking, but it was really pedestrian."

Bushnell's mind once again turned to vertical integration. He decided that the company's most valuable product was the map data Etak had synthesized for use in the devices. Reasoning that people with telephones received free Yellow Pages—a $9 billion advertising market at the time—why not sell Etak devices loaded with paid merchant info? "That was our core plan: sell the equipment and give away the mapping software loaded with McDonald's and Chevron Oil," he says. "If you wanted a list of five-star restaurants, a menu would come up showing the distance." Though Bushnell now laughs at the plan ("it shows how stupid you can be"), it was actually ahead of its time, anticipating not just applications like Yelp, but also all of today's geo-coded targeted marketing.

With its visual map interface, Etak had the basic look and feel of a modern GPS navigation system. It featured an arrow that always pointed in the direction of the destination, providing a rough sense of orientation for the driver. But it had no turn-by-turn directions. The technology existed—people had tinkered with algorithms that could compute driving directions since the early 1970s—but no one was sure how to optimize the technology. What if the system told a driver to go the wrong direction on a one-way street? How should these navigation systems communicate with drivers? Studies began to appear comparing methods to determine the best balance of information content, intelligibility, and driver safety. One Dutch survey compared the efficacy of moving maps with turn-by-turn directions and concluded that, with the exception of Etak, moving maps seemed to distract drivers, who generally made fewer errors with turn-by-turn systems.

Several researchers began to consider the problem in the context of cognitive maps. In the early 1980s, Benjamin Kuipers, a computer scientist, approached the idea from the standpoint of computational design rather than psychology. If a team were building a robot, he wondered, what would be the minimum requirements for the mechanism that gathered and stored information about the physical environment? At the most basic level, the robot would have to know how to combine "views" (sensory input provided by a specific location) with "actions" (a motor operation that changes the view) to create a route. A second level of knowledge would be the ability to assimilate a general conception of an environment's topography, where something is in relation to something else. The highest level would be a knowledge of exact distances and directions between the various parts of the topography. Kuipers argued that any cognitive map, whether a robot's or a person's, must contain these three levels.

In 1986, two geographers, David Mark and Matthew McGranaghan, adapted these three levels as a framework to consider what the ideal driver navigational system would look like. The lowest level, which they called procedural information, would be information presented in words, either written or spoken aloud. The second would take the form of a hastily sketched map, while the third would be represented by a professional map. They reasoned that anyone navigating a route *must* have procedural knowledge, and that if information were presented visually, such as on a map or computer screen, the driver would just be forced to perform the cognitive task of converting the information to procedural information. Since that takes time and effort, it would distract the driver and lead to navigational errors. Better to have procedural knowledge delivered another way.

To bolster their argument, they pointed to the results of a recent Bell Laboratories study that examined drivers' responses to different navigational aids. One group used a state road map; another used a customized road map that showed the route to be taken; a third received verbal instructions on a cassette tape that could be paused and rewound; and the final group used both the tape and the customized map. The drivers who made the fewest errors, drove the fewest miles, and reached their destination quickest were the ones in the tape-only group. The group that relied on the state road map fared the worst.

Perhaps this is why drivers today are so willing to cede all authority to that voice giving turn-by-turn GPS directions. Another clue may lie in a well-known psychology experiment first conducted in the late 1960s, in which subjects were shown a diagram depicting a large block-letter capital F, and told to commit the image to memory. Using their memory of the image, they were told to begin at the bottom left corner of the letter, and proceed clockwise, classifying each corner of the letter into two categories: points that were in the extreme top or bottom of the image, which received a "yes" response; and points that were somewhere in between, which received a "no." The way they were told to present these yes/no response varied. Some subjects were told to say them aloud. Others were instructed to tap their left hand to indicate yes and their right hand for no. A third group pointed to Ys or Ns on a sheet of paper. The results revealed that the pointing group was much slower than the other groups, requiring about twice as much time to communicate the responses.

The likely reason for the large discrepancy was the principle, observed by psychologists, that performing two cognitively similar tasks simultaneously is much more difficult than performing two cognitively different tasks. Scanning an image and pointing at something are both spatial endeavors. Responding verbally or by tapping are not spatial tasks, and are therefore easier to do while imagining the contours of the letter F. Another geographer, Scott Freundschuh, thought these results supported Mark and McGranaghan's conclusion. It is, of course, much easier—as well as safer—to drive while listening to a voice than to drive while attempting to read a map, but it might be easier on a cognitive level as well, since driving and interpreting maps are both spatial tasks, while processing spoken information is a verbal task.

The modern era of GPS car navigation began after Desert Storm, and no GPS startup exploited its potential better than a Kansas City-area company called Garmin. Gary Burrell and Min Kao were engineers at AlliedSignal, helping to develop a GPS receiver. (A prototype was on board the Rutan 76 Voyager, the record-setting aircraft that in 1986 became the first to circumnavigate the world with no stops or refueling.) After the project was discontinued, Burrell and Kao formed a new firm in 1989, its name a portmanteau of their given names.

With Magellan and Trimble at capacity filling military orders, Garmin found space in the market. As Trimble and Ashtech concentrated on high-precision GPS, Garmin's only major competitor in the low-end market was Magellan. Like Magellan, Garmin was able to design components that made the most of limited processing power, especially in the acquisition of satellite signals. Even at the end of the nineties, GPS devices designed for cars required the user to tediously download maps. Over the next few years, map data made directions more accurate, better processors made it easier and quicker for algorithms to compute turn-by-turn directions, and memory improvements finally put all the maps in the box. Garmin's C550 receivers, which hit the market in 2006, achieved full functionality. "Once solid-state memory had enough density to hold the entire country out of the box, that's when the market really took off," says Jay Dee Krull, one of Garmin's earliest hires.

By 2006, Garmin controlled 60 percent of the U.S. market for navigation equipment. Americans bought five million Garmin GPS receivers that year, as the company posted $1.68 billion in sales, a 64 percent increase from 2005. Fully half of the company's revenue came from car GPS units, with sales in that segment growing at an astounding 140 percent annually. "What Garmin has done is nothing short of invent an entirely new consumer electronics segment," raved *U.S. News and World Report.*

Even as Garmin amassed $900 million in cash and securities, with no long-term debt, it continued to be run as a lean outfit. As CEO, Kao paid himself a salary of $250,000 in 2006, and gave himself a bonus of $203. (Burrell retired from the company in 2004.) Today, Kao and Burrell are fixtures on *Forbes*'s annual ranking of the 400 wealthiest Americans, with a combined 2014 net worth exceeding $4.5 billion.

As Garmin enjoyed spectacular growth, the popularity of GPS navigation led to renewed interest in how these devices were affecting the behavior of users. But instead of just investigating navigation systems' design and effectiveness, some experts confronted the question of whether these new navigation systems might be weakening our cognitive map. Some of the widely praised attributes of navigation devices, especially their ability to present information in smoothly filtered ways that removed us from the bother of map-reading, came under closer scrutiny.

A 2006 German study, conducted by a group of psychologists and artificial intelligence experts, tested the hypothesis that users of a navigation system will remember less about an environment than those who use a map. Participants in the experiment walked a predetermined route through the zoo in the town of Saarbrücken carrying handheld computers connected by Bluetooth to a computer carried by an experimenter who followed several meters behind. He transmitted directions to the next segment of the route, beginning with a photo of the subject's current location. In addition to the directions, one group of subjects also received visual cues, such as a map and red line that appeared on the photo; others received only verbal instructions; and a third group received a combination of verbal and visual cues. A control group walked the route guided only by a crude map that showed the route with no landmarks, along with the same photos shown to the test subjects at the start of each segment.

Later, all participants were given a two-part test to gauge how well they remembered the route. They were asked to recall what directions they had taken—at what points they had turned left, right, and so on. The second part tested what the researchers called survey knowledge: "the spatial relationships between locations." The participants were shown thumbnail pictures of the intersections, and instructed to place them on a road map of the zoo. The researchers were essentially testing participants' ability to construct a cognitive strip map and a cognitive comprehensive map.

The data revealed a couple of interesting insights. First, while it seems obvious why map users would have better overall knowledge of the area they walked, why would they also have a better memory of the route itself? The researchers reasoned that the map users had to engage in more active learning, having to match the photos they were given of the route with the markings on the map. The researchers had predicted—wrongly, as it turned out—that the two test groups who received visual aids that provided spatial context would outscore the verbal-only group in the survey knowledge test. Instead, they concluded that because this information was not required for the act of wayfinding, the subjects were not forced to actively process it.

A similar study, conducted in Japan and published in 2008, involved actual GPS devices. Three groups of walkers were studied as they navigated routes in the city of Kashiwa. One group learned the route from direct experience, shown it by a guide who took them from start to finish, and then back via a circuitous route; the subjects were then instructed to walk from start to finish again, with no assistance. Another group was given GPS devices, with the complete route highlighted on the screen, while another was given a paper map with the beginning and end points marked, but no highlighted route. The results showed the GPS users exhibiting the weakest wayfinding acumen. They traveled at a slower speed and made more stops to reorient themselves than the walkers in the other two groups. They rated the overall task as more difficult than the group that learned the route by walking it. In post-walk tests, they had the lowest scores on memory of the configuration and topology of the route. The researchers concluded that the GPS system "was less effective than the maps and direct experience as support for smooth navigation."

A Cornell University study published the same year looked at GPS's effect on drivers, and reached similar conclusions regarding how GPS users "attend to objects in the paths they take toward their destination." The study "found evidence for loss of environmental engagement . . . the process of interpreting the world, adding value to it, and turning space into place is reduced to a certain extent and drivers remain detached from the indifferent environments that surround them." Their conclusion: "GPS eliminated much of the need to pay attention."

In the years when GPS auto navigation began to take off, brain experts were making important breakthroughs in the study of how spatial information is processed. Their findings suggest that there is a physical dimension to the cognitive map.

The first hint came from rat experiments performed in the early 1970s. John O'Keefe, a neuroscientist, reported that when rats were placed in certain areas of a room, cells in the hippocampus, which he labeled "place cells," were activated. More than thirty years later, two Norwegian neuroscientists, May-Britt Moser and Edvard Moser, expanded on O'Keefe's work. Their experiments demonstrated that when rats moved around the room, "grid cells" in a part of the brain linked to the hippocampus rearranged themselves based on the rat's environment. When a rat entered a new room, the cells lit up in a spatial pattern that corresponded to the location of the rat's head and the room's borders.

In 2010, a team at University College London confirmed the existence of grid cells in humans, forming a tidy geometric pattern. To conceive of how they behave, imagine walking into a room with a tiled floor. Some of the tiles, evenly spaced, light up when you step on them. "It is as if grid cells provide a cognitive map of space," Caswell Barry, a coauthor of the study, explained. "In fact, these cells are like latitude and longitude lines we're all familiar with on normal maps, but instead of using square grid lines it seems the brain uses triangles." (Further research revealed they are more like hexagons.) Other research at UCL has isolated two parts of the brain that help us as we navigate an environment, with one part noting the distance to the destination as the crow flies, and the other calculating the actual distance of the route.

If we do indeed have a kind of innate GPS, what happens to our brains as we transition into a world where these kinds of calculations are unnecessary, when GPS does it all for us? The short answer is we don't know yet, but there are some rumblings among cognitive experts to the effect that we may be undergoing fundamental changes. "Physical maps help us build cognitive maps," Julia Frankenstein of the Center for Cognitive Science at the University of Freiburg has argued. Frankenstein was lead author on an experiment done using residents of the German town of Tübingen. Wearing head-mounted displays, they navigated a 3-D virtual reality model of their hometown. At certain points along the way, they were asked to point in the direction of well-known landmarks that were not visible from the subjects' current perceived location. The results showed that the pointing was most accurate when the subjects were facing north. The greater the deviation from a northward orientation, the less accurate they were.

The participants were linking their perceived location to its position on city maps that they retrieved from memory—which, like most maps, were oriented to the north. The cognitive process the participants went through was, in a sense, more complicated then necessary, since they had spent a greater chunk of their lives navigating the city than looking at maps of it—and most of the locations they were asked to point out did not appear on maps of Tübingen. Some participants reported that they had not looked at map of the city for decades. The knowledge they had acquired just by navigating their city day after day was multisensory and tinged with memories of real experience, whereas a map is flat, and the only sense it appeals to is visual. Yet, when asked to organize the information they held, they still reflexively translated it into broad survey knowledge, a bird's-eye view. They willed themselves onto a map.

One important thing maps offer, the researchers noted, is stability. They aggregate information about the environment into one reference frame, and organize multiple navigational experiences into one reliable structure. Their two-dimensional models are a convenient catchall for our 3-D existence. "Our results support the popular belief that people have access to something like a map in their heads, and suggest that . . . this map is oriented north," the study concluded.

Other cognitive scientists have reached similar conclusions. Research has shown that when we are presented with an object, our interpretation of its shape depends on which part of the object is perceived to be the "top." If the object is rotated so that another part is in the "top" position, our perception of the object's shape can shift dramatically. This is probably because when we see what is in front of us, we seek out the direction of "up," since gravity is such a powerful reference axis.

Timothy McNamara and Christine Valiquette, both psychologists, have argued that something similar happens when we enter a new environment: "in effect, conceptual north is assigned to the layout, creating privileged directions in the environment." Unlike the vertical plane, so closely associated with gravity, the "ground plane," defined by the objects that surround us, has no such privileged direction, so we determine it based on our perspective. When we recall the environment later, "the dominant cue is egocentric experience"—that is, we assign a new "north."

In one experiment cited by McNamara and Valiquette, subjects were told to stand in a cylindrical room and study the objects in it from three assigned perspectives. Later, in another room, they were asked to point to the directions of objects in the first room, as seen from various perspectives. For example, a subject might be instructed to imagine standing in front of the clock, and then told to point in the direction of the book. "The surprising result was that only the first study view . . . appeared to be mentally represented," McNamara and Valiquette write. "Pointing judgments were quite accurate for the imagined heading parallel to the first study view but no more accurate for the second and third study views than for novel headings."

Spending our days moving through various environments, we fill in the details of our cognitive map based on our egocentric experiences. Can the granular detail of that map fade through misuse? "The problem with GPS systems is, in my eyes, that we are not forced to remember or process the information—as it is permanently 'at hand,' we need not think or decide ourselves," Frankenstein says. "The more we rely on technology to find our way, the less we build up our cognitive maps." Life becomes a series of strip maps: "we see the way from A to Z, but we don't see the landmarks along the way . . . developing a cognitive map from this reduced information is a bit like trying to get an entire musical piece from a few notes."

Moreover, this suggests that we absolve ourselves from even having egocentric experiences to build upon. In some general sense, we lack reference points, stable spots that anchor our position in the world. Without these authoritative positions that, in a very real sense, add meaning to our world, we are left floating. Perhaps there is something to the explanation, by those who have driven their cars into rivers and over cliffs, that GPS told them to do it.

The next frontier of research would be to investigate whether GPS use can cause physiological changes. A British study, published in 2006, made headlines by revealing that the brains of London taxi drivers, whose licensing requires that they demonstrate recall of 25,000 city streets, plus the locations of landmarks and points of interest, contain more gray matter in the region of the hippocampus responsible for complex spatial representation than the brains of London bus drivers. Brain scan results from retired taxi drivers suggested, without being conclusive, that the volume of gray matter decreases when this ability is no longer required.

The idea that something similar may be happening on a large scale, as GPS use becomes more ubiquitous, is plausible enough to be taken seriously. More recent research has demonstrated that someone who does intensive exercises to improve navigational skills can exhibit changes in the hippocampus. "Based on that, I think it's possible that if you went to someone doing a lot of active navigation, but just relying on GPS—the assumption being that they'd be minimizing the brain's use for navigation—you'd actually get a reduction in that area," says Hugo Spiers, one of the scientists who conducted the London cab driver study. "I would love to know," he adds, but cautions that conducting a rigorous study with human subjects, controlling for all the variables that affect brain function, would be extremely difficult and probably prohibitively expensive.

Meanwhile, in the absence of viable human subjects, the rodents still run the mazes, exploring the mystery of the stimulus-response/cognitive map dichotomy elucidated by Tolman more than a half century ago. In one experiment described by Alex Hutchinson in his *Walrus* article, some mice were trained to run a maze that encouraged them to use stimulus-response strategies, while a second group ran a version of the maze that forced them to use their cognitive map. After just five days, MRIs showed that the second group exhibited an increase in gray matter volume. Dissection revealed that the increased volume was caused by a growth in the number of connections between the neurons.

Even if we were to discover that GPS use is reshaping the physical contours of our brains, there would still be an elusive, unknowable quality to death by GPS. The presence of grid cells notwithstanding, our personal cognitive map is not reducible to gray matter. It remains locked away in the black box of the nervous system, accessible when we need it but never fully unfurling for us to examine. Like Tolman said, maps are pictures of reality, not replicas. Our own personal cognitive map is the prism through which we glimpse that reality. How can we even tell if it is narrowing into a strip map? We have no other means of perception.

Even if we recognize a narrowing map as the only explanation for someone's death by GPS, what have we really learned? In March 2015, Iftikhar and Zohra Hussain were driving from their home in Chicago to Indiana to visit family. As Iftikhar approached a bridge that spanned the Indiana Harbor and Ship Canal, he ignored orange cones, "Road Closed" signs, and other deterrents meant to keep cars away, since the bridge had been closed for repairs since 2009. The car plunged off the bridge, dropping nearly 40 feet to the ground near the water. Iftikhar managed to escape before it burst into flames. Zohra died from her burns.

A local paper, citing a police investigator, reported that Iftikhar "was apparently paying more attention to the navigation system than what was in front of him." Assuming that were true, the question still remains: What was going through his head?

Nearly two months after the Chretiens disappeared, three hunters in an all-terrain vehicle, somewhere in the Independence Mountains, came across a Chevy Astro. The three—a husband and wife, and the woman's father—cautiously approached. From inside the vehicle, a woman wearing a plaid shirt and jeans managed with great effort to open the sliding door and poke her head out. The older man flashed her a friendly "A-OK" symbol by joining his thumb and forefinger in a circle. Rita Chretien shook her head, barely able to rasp the words, "No, I'm not OK."

The Chretiens had remained on the road that night, eventually realizing they had no choice but to press on. The road was too narrow and treacherous for them to turn around. At an elevation of 6,000 feet, the Astro had slid into a gully and gotten stuck in the mud. In the morning, they discovered that the road soon narrowed into a trail. It still looked to them like they were heading in the direction of Mountain City, which they estimated was about 27 miles away. So they began walking, venturing out a few miles and then returning to the van as night approached, the temperature dropped, and the rain arrived.

The next morning, they decided that Albert would set off on foot again. Rita had injured her knee on the hike the day before and found it hard to walk. Albert figured it would take him between two and three days to reach Mountain City. Dividing their meager supplies, they decided that Al should take the bag of chocolate-covered almonds for energy. Rita's take included a small sandwich bag filled with trail mix, some hard candy, and fish oil. Al wrote down the GPS coordinates for Mountain City, and took the Magellan with him.

Rita carefully rationed her food, eating as little as she could each day. She melted snow and gathered water from a nearby creek. She passed the days praying, meditating, writing in her journal, reading books and her Bible, and sleeping as much as possible to conserve energy. She wrote notes in case she was found: "Please help. Stuck." "We're headed to Vegas. Got lost." "No food. No gas. . . . Al went to get help. Find Mountain City. Did not return! . . . Maybe died along way?" One note gave her present GPS coordinates. She grew too weak to walk to the stream, and drank what water she could from puddles. Just before she was rescued, she decided she had one more day left in her. She put on fresh socks, wrapped a blanket around her, and prepared to die.

Her rescuers gave her what little food and water they had, but realized she was too weak to ride on an ATV. One remembered a ranch eight miles away, and they asked if she could hold out for one more hour. They found the ranch, called 911, and led the sheriff's chopper to Rita's location. By the time they reached her, she had torn down the notes and packed her bags. She was smiling, and the rescuers swore she'd fixed up her hair. She was airlifted to a hospital, where she was gradually reintroduced to food and spent Mother's Day with her children. Rita Chretien, fifty-six years old, with no outdoor experience and next to no provisions, had somehow survived a trial that would have taxed the most hardened survivalist.

In December 2012, nearly two years after her ordeal, Rita and four of her friends took another road journey, what she called a "trip of gratitude." She wanted to visit the regions where people had organized searches for the Chretiens, to meet as many of them as possible and say thank you. She also had a chance to see her rescuers again. They had been back to the site where they found her several times, trying to find where Albert had gone. Now they wanted to take her back there, too. She was initially dubious, but accepted the invitation. Little had changed. The Astro's tracks were still visible. "I showed them where I got my water," she says today. "It was very emotional, seeing my old fire pit."

She was also able to meet the man from the Elko County sheriff's office who had organized the search party in the area. "I had tried to figure out how on earth we got lost," she says. "He said he realized we had followed exactly what the GPS said, because he went and followed what I told him, and from there ended up exactly where we ended up."

Exactly one week after returning from her gratitude trip, she received a call from the sheriff. Albert's remains had been found some seven miles from the van. He had made it a little more than halfway to Mountain City before succumbing to hypothermia and exhaustion. The Magellan, designed to run off of a car's battery power, had probably fizzled soon after he began his journey.

Rita remains remarkably serene and philosophical regarding her experience. "I'm not so sure I want to venture out on strange roads anymore," she says, laughing quietly. "I just stick to the main roads now." In 2015, she remarried.

In his final hours, Albert's course had veered north. He ascended 2,600 vertical feet, through snowdrifts taller than he was. "He did a lot of unnecessary climbing," a sheriff's deputy noted. "He was heading literally for the summit of the mountain."

Rita thinks she knows why. "I believe Albert climbed toward the peak to find shelter, but also to have a good look around," she says. "To see where to head from there."

CHAPTER SIX

The Hornet's Nest

Few capital cities are as detached from their polity as Juneau, Alaska. No roads connect it to the rest of the state or neighboring British Columbia. Although it is accessible by sea, via the Gulf of Alaska, air travel is this city's lifeline.

Most flights into Juneau are on Boeing 737s operated by Alaska Airlines. A pilot approaches the airport from the south, following the Gastineau Channel, a narrow waterway that separates Douglas Island from the mainland. As the plane descends, it drops below the peaks of the Chilkat Range, less than a mile away on both sides. The pilot does not make visual contact with the airport until about ten seconds before touching down. The plane banks right, makes a hard left turn, and the oversize "26" painted at the foot of the airport's major runway comes into view. One final quick descent, and the plane is on the ground.

Until the early 1990s, the process of landing at Juneau had remained the same for decades. While maneuvering through the channel, a plane followed a radio beacon from the airport. But because the plane was flying lower than the mountains that formed a barrier between it and the airport, the beacon could not be placed on the runway. It beamed its signal from a spot at the end of the channel, about a 15-degree remove from the runway. Pilots aimed for the beacon and performed a final course correction left before landing. Flight rules stipulated that this correction could not occur until the pilot had visual verification that the plane had passed a specific notch in a ridge—even at night or when flying through low cloud cover, which was often. The last thing a pilot wanted to see (the last thing a pilot *would* see) was a snowy peak straight ahead, gleaming in the headlamps.

At that point in the descent, the plane was required to have a *minimum* altitude of 500 feet—more than enough vertical space to protect the plane from obstacles. But measuring altitude before the 1990s was an inexact science. Altimeters functioned by comparing the barometric pressure outside the plane to a baseline figure provided by the airport, a method that could lead to calculations that were off by several feet. In 1971, an Alaska Airlines flight en route from Anchorage collided with a mountain on approach, killing all 111 people on board; at the time, it was the deadliest single-plane accident in U.S. history. That accident loomed large for flight crews, especially in those inevitable nighttime descents when the plane was flying at a lower altitude than a skyscraper. "Particularly at night, I could feel the stress and a trickle of sweat down my armpits," Steve Fulton, a veteran Alaska Airlines pilot, recalled.

Fulton, who had an engineering background, envisioned a "highway in the sky" that could make the Juneau approach simpler and safer. It would involve GPS coordinates creating a virtual lane that would help pilots navigate that stressful landing. As long as pilots followed those coordinates, they could be confident they would be safe. Fulton left the airline to launch Naverus, a company that adapted a technology called Receiver Autonomous Integrity Monitoring, recently developed by the Stanford GPS Laboratory and Mitre, a research organization that manages many federally-funded tech projects. A RAIM-equipped GPS receiver required a signal from six GPS satellites, instead of the standard four. As with any GPS reading, three satellites determine latitude, longitude, and altitude, while a fourth resolves timing ambiguities caused by the receiver. A fifth satellite reviews these calculations and determines any faults. If it finds any, the sixth satellite determines which of the four satellites caused the error. The system then makes the necessary corrections.

Alaska Airlines introduced RAIM for Juneau approaches in 1996, and averaged about one RAIM-enabled landing every day for the next fifteen years—more than 5,600 flights. RAIM was not just a safety feature, a way to calm pilots' nerves; it was also an efficiency booster. Without GPS, atmospheric conditions or poor visibility would have forced the airline to cancel one of those flights every week, costing the airline about $1 million annually. By 2011, Alaska Airlines was using RAIM at thirty airports. In that year alone, RAIM prevented the diversion of 1,500 flights, decreased fuel consumption by 210,000 gallons, and saved the company between $15 million and $19 million.

RAIM was an aviation milestone, the first GPS-powered automated navigation system sanctioned by the Federal Aviation Administration. From the early days of GPS, Brad Parkinson had envisioned commercial aviation as the most obvious nonmilitary application of GPS. One goal of his Stanford GPS Laboratory, formed in 1984, was to develop these applications. Reagan's announcement, following the Korean Airlines disaster in 1983, that GPS was available to all, specifically mentioned civil aviation. But the FAA initially showed little interest in integrating GPS into standard aviation procedures. Outside the U.S., resistance was even stronger. Developing nations, Communist bloc countries, and even a few Western democracies did not relish the idea of their aviation infrastructure depending on a system controlled by the Pentagon.

The agency's reticence was understandable. Like the fledgling commercial GPS industry, it was reckoning with an unproven technology and a satellite constellation that was still incomplete. "When we were in the general aviation business, they were very frustrating, until I actually understood the problem,"

Charlie Trimble says. "The FAA's job is to keep things safe, and historically they have been successful at that by slowing the rate of change. The FAA fulfills its mission by wrapping policies and procedures around obsolete technologies."

GPS had a high bar to clear to meet the FAA's safety requirements. For a navigation system to be certified for precision approaches—when a pilot relies solely on a plane's instrument readings, without visual confirmation—it must demonstrate, through rigorous testing, a success rate of 99.9999999 percent. Seven 9s, no fewer. That means that once every hour there is a one in a billion chance the system will be unavailable or transmitting faulty data. The maximum success GPS alone is capable of is 99.99999 percent. Five 9s, not seven—a failure rate measurable in millions, not billions. That's not good enough—especially for busy airports. "Five nines," Trimble says, "gives you a near-miss over O'Hare once a day."

RAIM was certified for lateral positioning but not for altitude, disqualifying it as a tool for precision landings. Its measly five 9s was not the problem. The next major GPS aviation project pursued those two 9s. In 1994, the FAA began work on the Wide Area Augmentation System. For technical reasons, WAAS is not considered a differential GPS system, but that is basically what it is. WAAS uses three dedicated satellites at an elevation twice as high as the GPS satellites, in a formation called a geostationary orbit, which means they rotate with the Earth, each always over the same geographic point.

Using its own dedicated infrastructure, WAAS acts as a sort of independent overseer of GPS. WAAS monitoring stations, spread across the continent track the GPS satellites and compute errors and corrections. This data is sent to three master control stations, in Los Angeles, Atlanta, and Washington DC, which process it and relay it to the WAAS satellites, which in turn relay the corrections to anyone with a WAAS-compatible GPS receiver. The satellites also function as their own independent GPS-like system, sending pulses that the receivers use to make ranging calculations. On top of all that, WAAS constantly analyzes its own GPS corrections, estimating how much error still remains.

All of that to land a plane without looking. Even before it could land planes—when the system was up and running, but was not yet certified for aviation—beet farmers and other precision agriculture practitioners added WAAS to their arsenal of satellite positioning tools. When Congress held hearings to determine if the project should receive continued funding, they showed up, irate. "It was the farmers who shouted, 'Don't you dare touch that system,'" says Per Enge, a Stanford professor and one of the major designers of WAAS. "They were loud, very loud."

It took the government nine years to complete WAAS. Any aircraft in North America with a WAAS receiver could now make a precision landing using GPS. "Those protection levels are guaranteed to seven 9s," says Tom McHugh, the FAA's technical director for WAAS. "You get to sue the FAA if they're wrong."

Other wide area systems similar to WAAS emerged, such as European Geostationary Navigation Overlay Service (EGNOS) and the MTSAT Satellite Augmentation System (MSAS) in Asia. And GPS continued to infiltrate aviation. In 2012, the FAA launched NextGen, an ambitious project to renovate the air traffic control infrastructure so that the primary technology is GPS, rather than radar. Radar's lag time (a plane might be miles away from the radar reading) forces airports to adopt very conservative landing procedures, funneling all aircraft through the same approach routes, like a crowd queuing for a single escalator. A plane approaching San Francisco's coastal airport currently flies 50 miles past it, makes a U-turn near the city of Livermore, and heads back to join what Enge calls "the conga line over the South Bay." Separated by a minimum distance of 2,000 feet, planes wait for their turn to descend. If this were a smaller airport, the conga line would be more like dancing the limbo, as planes descend in a sort of stair-step formation: dropping altitude, flying straight for a while, dropping again, repeating the process until the final approach.

GPS gives pilots more leeway in plotting an approach to the airport, with gradual descents and less stair-stepping, and a reduced minimum space between planes. Basically, it allows crowded airspace to be used more efficiently, and these little changes add up. Every time a Boeing 747 lands in San Francisco without taking a side trip to Livermore, 1,600 gallons of fuel are saved. The FAA estimates that the planes vying for space around Washington, DC's two airports will save 2.3 million gallons of fuel per year, and reduce emissions by 7,300 metric tons. Commercial airlines aren't the only ones reaping the benefits. In Memphis, Federal Express can add nine flight operations per hour, with annual savings of almost $22 million. Residents of Louisville, Kentucky, can breathe easier, knowing that arrivals at United Parcel Service's central processing facility are burning 7,761 fewer gallons of fuel. In a world of low margins, scarce resources, and race-to-the-bottom competition, GPS is a powerful means to chip away at costs.

When Alaska Airlines began to use GPS in 1996, it was a watershed moment for the aviation industry and also for GPS, heralding its integration into what governments and security specialists call the critical infrastructure—an inclusive term that refers to the systems, installations, and industries that make modern life possible. The U.S. Department of Homeland Security calls the critical infrastructure "the backbone of our nation's economy, security, and health . . . so vital to the United States that their incapacitation or destruction would have a debilitating effect."

The U.S. officially designates sixteen critical infrastructure sectors, including energy, financial services, dams, and food and agriculture. All but three of them utilize GPS for some essential functions. For many countries around the world, the same ratio applies. GPS remains the world's only global utility, not merely because it is free. Unlike other utilities such as gas, electricity, and water, GPS is renewable—the amount of GPS receivers in the world could double tomorrow without the system skipping a highly synchronized beat. No running out of bandwidth, or phone numbers, or IP addresses. We can use all the GPS we want, and it doesn't cost us a cent.

For now, and for the foreseeable future, GPS is the world's global navigational satellite system. The only comparable system in the world, Russia's GLONASS, is a distant second to GPS, not nearly as robust or reliable. The European Union's unfinished Galileo project has been plagued by delays and setbacks. GPS is the only critical infrastructural system on Earth that transcends national borders. All countries have a stake in it.

America's anxiety about the security of GPS is therefore the world's anxiety. And lately, that anxiety has increased. For the first time, in 2010, the National Space Policy articulated America's official position on GPS, which included a proviso that other countries' GPS-like systems "may be used to augment and strengthen GPS," and another that pledged U.S. support for "international activities to detect, mitigate, and increase resiliency to harmful interference to GPS." The fear is not a catastrophic global failure of GPS, a complete cessation of signals. The satellites themselves would be nearly impossible for someone to damage. The worry is localized disruptions, especially attacks on GPS-dependent systems that would have a cascading catastrophic effect.

In the U.S., the beginning of that anxiety coincided almost exactly with GPS's debut in the skies over Juneau. President Clinton convened a Commission on Critical Infrastructure that examined all the segments and concluded that the largest potential vulnerabilities involved GPS and aviation. More generally, the commission was concerned about potential weaknesses in the transportation sector.

The intersection of the two—GPS use in transportation—is a useful introduction to the deeply embedded nature of GPS.

Its role in guiding commercial jets is merely the most dramatic example. GPS insinuates itself into the various ways people move through any major city. Nearly ten years ago, the head of the Chicago Transit Authority stated that GPS "is quietly permeating the infrastructure." Today, the infrastructure is saturated. In New York, as in many other urban areas, smartphone apps allow riders to track the progress of approaching city buses. The agency that oversees the bus system also uses GPS to regulate the system. Behind the scenes, in various secure rooms around the city, where journalists are not welcome, managers stare at screens to track the more crowded routes—they may instruct a bus driver to skip the next stop, for example, if another bus is gaining on it. Once a day, the GPS time signal synchronizes the city's stoplights—especially important in areas where the signals are timed to maximize flow. Some buses even exert some control over those signals—if a centralized computer decides that a red light will put the bus behind schedule, it will keep the light green long enough for the bus to proceed.

Across the world, cargo ships use GPS, both on the open ocean and when moving in and out of port. Their merchandise is tracked by GPS as it moves around the docks. GPS guides ambulances and fire trucks, dispatched by operators whose calls are time-stamped by GPS. On the railways, nearly 3 million freight cars are equipped with GPS equipment. The burgeoning cottage industry of "positive train control"—GPS-based information systems that control rail traffic—will gain traction in the coming years.

And we haven't even mentioned drones.

That's just a thumbnail sketch of the transportation sector. Scratch the surface of most of the critical infrastructure, and you'll find layers of GPS. The penetration of GPS has been swift and stealthy. It is as though the world woke up one day to discover that this technology, built to drop bombs, now has its tentacles everywhere. A few voices worry that all this has happened without enough thought being given to the system's security.

The voices are getting louder—and while they are concerned about people disrupting GPS by jamming its signals, this is by no means their biggest fear.

Until it disbanded in 2014, an interdisciplinary think tank of creative destruction occupied a small barracks-like building on the idyllic campus of the Argonne National Laboratory, outside Chicago. The job of the Vulnerability Assessment Team was to break the unbreakable. The work they performed fell under the somewhat obscure rubric of "physical security." Experts in this field probe the weaknesses of methods and devices that protect physical assets (buildings, people, etc.), as well the physical methods, such as fences and access control devices, that safeguard digital data, intellectual property, and other virtual assets. Among its conquests, the Argonne team demonstrated the ease with which one can fool electronic pass card readers, biometric ID scanners, and computerized voting machines—and also tackled more quotidian forms of security, from handcuffs (surprisingly easy to slip out of) and shipping crates that supposedly cannot be opened, burgled, and reshut without breaking a telltale plastic tab (they can).

In the early 2000s, when the Vulnerability Assessment Team worked out of Los Alamos National Laboratory in New Mexico, the team leader, Roger Johnston, found himself thinking about nuclear waste. The Waste Isolation Pilot Plant, near Carlsbad, in the far south of the state, is one of the few sites worldwide where nuclear waste is buried deep underground—and is America's only repository for transuranic waste, much of it from old Cold War nuclear production facilities. Although Johnston did not know the exact security measures in place for transporting radioactive material to WIPP, he knew one level involved using GPS to track the vehicles. If someone wanted to hijack a truck—to obtain material for a dirty bomb, perhaps—how easily could they hack the GPS protection?

A potential terrorist could jam the GPS signal so that the trucks disappeared from the monitors: one form of potential sabotage, but not the worst. Johnston was more concerned about hijackers taking control of a truck while simultaneously broadcasting a mock GPS signal that overwhelmed the real one. This "spoof" signal could generate coordinates that made the monitors believe the truck was on course. The hijackers could be many miles away before anyone realized something was wrong.

Johnston considers it axiomatic that no lock in the world is unbreakable. For ideal security measures, he prefers the model of a seal, something that does not necessarily prevent a break-in but leaves incontrovertible proof of a breach. That was what the plastic tabs on shipping containers were supposed to do, but did not. The Vulnerability Assessment Team's alternative measures include the use of stickers, embossed with a sparkly pattern, that utilize our eyes' amazing ability to spot differences. The shipper places the sticker over the container's opening, and photographs it. The receiver also photographs the sticker. If a thief has managed to remove the seal without breaking it, and then replaced it, its position will be *slightly* different, a discrepancy revealed by overlaying the two photos. The stickers do not prevent tampering—but compared with plastic tabs, they make it much more difficult to hide the evidence of tampering.

The Vulnerability Assessment Team decided to test the porosity of GPS by attempting to spoof it. They began by renting, for $1,000 per week, a GPS satellite simulator, a legal device that tests the accuracy of a GPS receiver by simulating the active GPS satellite constellation. The simulator connects directly to the receiver, so its signal does not go out over the air—but it didn't take much tinkering to attach an antenna to the simulator so that it was broadcasting a weak signal. To boost it, the team spent another $300 on a GPS signal amplifier. They placed both components in the back of flatbed truck, along with a desktop computer, monitor, and battery. On the truck's grill, they mounted an antenna.

After enacting various hijacking scenarios, the good news, the team discovered, was that this kind of attack would require very precise execution. The hijacker would have to first jam the real GPS signal in the vicinity of the target truck, while seamlessly introducing the bogus signal, all while remaining no more than 15 feet from the target—not impossible to imagine on a highway. After control had been gained over the target truck, the attack truck would need to maintain a distance from it of no more than 30 feet, feeding the spoofed signal to the truck's GPS receiver so that it could continue to accept the fake coordinates and communicate them to the monitoring center. This logistical restraint made the hijacking a difficult proposition.

The actual spoofing, however, was shockingly easy. A terrorist or technology thief would require about $2,000 of legal equipment to disable the Defense Department's billion-dollar signal.

Johnston continued to organize GPS spoofing experiments over the next few years, always using over-the-counter components. By around 2005, the team had figured out how to spoof a GPS receiver from nearly two miles away—feeding it bogus location and speed data. Video footage of experiments conducted two years later shows three stationary GPS receivers mounted on a tripod, the camera zooming in to document that the receivers are all displaying an estimated speed of 600 miles per hour. In separate experiments, the team proved that they could spoof a timekeeping device that was getting a time feed from GPS. From more than 300 feet away, they radiated a bogus signal in the direction of an antenna attached to a clock connected to a laptop computer, with the time displayed in huge letters across the screen. At one point the display suddenly changed from 3:00 (the correct time) to 11:58.

"Current GPS receivers are relatively stupid," Johnston and a colleague concluded in 2003, reporting their findings. The most glaring example: a receiver would happily accept a fake signal that was thousands of times stronger than the fragile GPS signal.

The holes were so glaring, in fact, that plugging them seemed relatively easy. The Vulnerability Assessment Team concluded that a few software tweaks could be the sparkly stickers that teach GPS receivers to spot a crudely spoofed signal. Many years later, Johnston still maintains that any existing GPS receiver can be retrofitted for no more than $20. But when Johnston and his team presented their findings at conferences and in security journals, they were met with indifference or worse. Pentagon officials, he says, nitpicked at the team's methodologies. "We were ignored and then we were ridiculed—a lot," he says. "We've kind of given up, because we got tired of people either not listening to us or laughing at us."

Johnston still thinks about GPS. The problem, as he sees it, is not that spoofing is possible—no lock is unpickable—it's that spoofing is such a snap to pull off. "It's one thing to stop James Bond," he likes to say. "When Grandma can spoof GPS, we've got a huge problem."

Eventually—inevitably—someone decided to be James Bond.

Around the time the Vulnerability Assessment Team was taking an interest in GPS spoofing, Todd Humphreys was a young Cornell engineering graduate student. His area of concentration was software-defined radio, the processing of radio signals via software (as opposed to hardware components). Because Humphreys had applied these principles to building GPS receivers, an article by Logan Scott in a GPS trade magazine caught his attention. In 2001, the U.S. Department of Transportation and the Pentagon had conducted a joint study of GPS vulnerabilities. The results were released in a document known within GPS circles as the Volpe Report, which warned that as GPS became a bigger part of the critical infrastructure, it would form a "tempting target" for "individuals, groups, or countries hostile to the U.S." The Volpe Report was the first official acknowledgment that spoofing posed as much of a threat to GPS as jamming. Scott warned that the findings of the Volpe Report were being ignored, while society's dependence on GPS was increasing.

What struck Humphreys was that the worst-case-scenario spoofing attack that Scott described was something Humphreys was pretty sure he could pull off. Humphreys had a background in magic, having performed tricks at children's parties to make money as an undergrad. What was spoofing if not an elaborate sleight-of-hand? Intrigued, Humphreys sought out one of Roger Johnston's articles on VAT's spoofing experiments, and was shocked by their simplicity. From his perspective, this spoofing was so obvious that it would be detected almost immediately.

"There was an emerging consensus that this wasn't a big problem, because with some fairly simple techniques you could defend against spoofing," Humphreys says. "But I read through those techniques and said to myself, 'Wow, they're way overconfident here, with a software-defined radio there's a way to defeat every one of them.' I realized that people who were offering techniques had not tried them or built a spoofing device." Humphreys decided it would be a good idea to build a real spoofer, "so we wouldn't be so unhinged from reality." A successful spoofing attack would require a very sophisticated device—something strong enough to overpower GPS, cunning enough to imitate it, and subtle enough to avoid detection. It would be something that, Humphreys was fairly certain, had never been built.

After convincing his advisor, and some vetting from the school's lawyers, Humphreys began the project. He did most of the theoretical work at Cornell, and began the physical construction after he accepted a faculty position at the University of Texas at Austin—in his spare bedroom at first, and then, with the help of grad students, at the school's Radionavigation Laboratory, which he oversees. "We decided the Volpe Report had been a warning cry in 2001, and by 2008, when we finally published our work, that was enough time," Humphreys says. "The community had had fair warning, and we were now going to increase the temperature to get our point across."

A few years later, in 2011, Iranian military forces near the city of Kashmar captured a drone aircraft (also known as an unmanned aerial vehicle, or UAV) operated by the CIA. One theory, which Humphreys found plausible, was that Iran had spoofed the drone's GPS receiver. A few months later, Congress directed the FAA to finalize rules regulating the operation of civilian drones in the U.S. Thousands of devices were poised to enter the airspace, with organizations planning to use them for everything from search and rescue and weather forecasting to monitoring ports and burrito-and-beer delivery. Most drones do not rely solely on GPS to navigate, but GPS is "the bulwark," Humphreys says, "the one the other sensors defer to."

The Iran incident and the congressional directive convinced Humphreys that it was time to "kick the tires and carry out a demo." As it happened, the Department of Homeland Security was soliciting proposals for public demonstrations of GPS vulnerabilities. Humphreys offered to spoof a commercial drone—which is how he and his team found themselves in June 2012 at GPS's original proving grounds, the White Sands Missile Range in New Mexico, joined by an anxious group of government and military officials.

There was a nontrivial possibility that Humphreys and his team could end up looking like idiots. Their experiments in the lab had demonstrated their spoofer's ability to bewitch a drone, but testing it outside on a real drone would have required them to break the law by interfering with GPS. The closest they'd come to a full dress rehearsal was a trial run in UT's stadium—"they even moved football practice for us, which was a surprise," Humphreys says—at which the spoofing signals were fed to a hovering UAV via lightweight cables. The drone they were using for the test was a Hornet Mini, a helicopter weighing between eight and ten pounds and retailing for $80,000, popular with law enforcement, made by a Georgia company called Adaptive Flight Inc. One of Humphreys's students would fly the Hornet, another would operate the spoofer, and personnel from the Central Inertial and GPS Test Facility, the group tasked with evaluating all GPS components for the Department of Defense, would oversee the trial.

The students perched on a hilltop about a kilometer away from the test facility, where Humphreys nervously waited for the show to begin. The pilot manually guided the Hornet to an altitude of 40 feet, commanded it to hover, and switched over to autopilot mode. The drone was now governed solely by its sensors. As the Hornet maintained its position, the other student gradually unleashed the spoofed signal to do its seduction dance. It whispered in the drone's ear, telling the Hornet that it was beginning to drift upward. The drone had its orders—*don't deviate from 40 feet in the air*—so it took immediate action. If it had moved up, it better get back down, so it began a quick descent, heading straight for the desert floor. Standing next to Humphreys was an ex-Blackhawk pilot who had been recruited for just this contingency. With "steady nerves and a quick reaction time," Humphreys recalls, the man used controls to wrest the Hornet away from its demons. Like a sport fisherman landing a bass, he managed to yank the Hornet back up before it could crash. "Up to the very last second, we were prepared for it to be a failure," Humphreys says. "Thankfully, our math worked out. It was beautiful."

A month later, Humphreys flew to Washington to testify in front of the House Committee on Homeland Security. Humphreys made a good witness. Clean-cut, handsome, earnest in the manner of someone who dashes into phone booths and for some mysterious reason is never around when the caped crusader arrives, Humphreys excels at communicating his dire warnings so they don't sound like an academic hobbyhorse. The good news, he said, is that there were, at most, one hundred people in the world who could build a spoofer as powerful as his. For the moment, it was probably beyond the capabilities of organized crime or terrorist groups, but "well within the capability of near-peer nation-states."

Nevertheless, the ingenuity of spoofers seems poised to collide with the imbecility of some drone aficionados. In the summer of 2015, private drones flying near wildfires in California forced authorities to ground firefighting aircraft. Around the same time, at the annual DEF CON hacker convention in the summer of 2015, two researchers from an elite team within Qihoo 360, a Beijing-based Internet security company, demonstrated an inexpensive GPS spoofer they had built using commercially available components. The news was not so much that they had created a spoofer for $5,000—but that they posted their code on the online repository GitHub, for anyone to download. At least one person—someone affiliated with England's University of Bath—has experimented with it, concluding that, after a few modifications, the code works.

During his testimony, Humphreys emphasized that the takeaway from his experiment is not that drones are vulnerable. (As the firefighting-thwarting UAVs demonstrated, drones don't need to be hacked to be meddlesome.) Spoofing a drone "is but one expression of a larger problem," Humphreys said. "Insecure civil GPS technology has over the last two decades been absorbed deeply into the critical systems of our national infrastructure."

Dig deep and you discover a GPS you probably did not know existed.

Measured by total usage, GPS's biggest asset is not its pinpoint positioning—the descendant of Brad Parkinson's five-bombs-one-hole ethic. It dates back further, to Roger Easton's fascination with Timation's time-transfer experiments. You and I use GPS to know where we are, but *the world* uses it to regulate time. First and foremost, GPS is a clock. As a timekeeper, GPS is impeccable—always available, always accurate, always free.

The world-designated "official" time standard is Coordinated Universal Time (UTC), a more intricate version of the previous standard, Greenwich Mean Time. The International Bureau of Weights and Measures (BIPM), located in a suburb of Paris, manages UTC by averaging data from 200 atomic clocks in national laboratories around the world. One of them is the Master Clock at the U.S. Naval Observatory, the source of GPS time. UTC's basic unit of time is the SI second, an atomic standard based on the behavior of caesium atoms.

However, UTC also conforms to mean solar time, the preatomic clock conception of time measured by Earth's rotation. Because the planet's rotation is gradually slowing (mostly due to the effects of tides), making our days longer, every few years an additional second (leap second) is added to UTC to keep it in line with solar time. GPS time is based solely on an atomic standard, so the leap seconds do not apply to it. The GPS "epoch"—the dawning of time, as GPS now knows it—began at the stroke of midnight, when January 5, 1980, clicked over to January 6. Because of leap seconds, GPS time is now several seconds ahead of UTC. Embedded in the GPS signal is information about this time discrepancy, so that a GPS-enabled device, such as a phone, displays time that is more or less consistent with UTC.

GPS is not so much the world's clock as it is the world's most powerful stopwatch, a perfect way to *manage* time. In this way, it does play a key role in determining UTC. Timing laboratories routinely compare their clocks' output with other atomic clocks around the world. In the pre-GPS 1970s, Judah Levine designed the clock at the National Institute of Standards and Technology (then called the National Bureau of Standards), in Boulder, Colorado. Levine, who still heads NIST's frequency and time division, would periodically drag two battery-powered version of the NIST clock halfway around the world for synchronization at the

Paris bureau. "We'd get on the airplane with the two suitcases, and we'd put them in the overhead thing," he says. "We would get off the plane at Dulles Airport, and the Naval Observatory guy would meet us with his own portable clock, and we'd compare our time with his. Then we'd get back on the plane and reach Paris the next morning. We'd be met by folks who would take us to the timing laboratories. We'd compare clocks and then reverse the trip—carry 'em back."

GPS satellites, with their unerring pulse, offered an alternative, even with an incomplete satellite constellation. All that was required to compare two geographically distant clocks was one satellite in a line-of-sight position for both. This is called the "common-view" method. Both laboratories agreed to observe the same GPS signal (for example, one leaving the satellite at 08:45:15, based on GPS time), and note its arrival time. The actual comparison required complex calculations involving dial-up modems and timesharing computers, but the results—readings that were anywhere from 10 percent to 100 percent more precise— spoke for themselves. "Once GPS became more or less available, that was it," Levine says. "It was so much better that we never went back to the greasy kids' stuff."

The GPS common-view technique works so well that scientists use it to track neutrinos, the smallest subatomic particles in the universe. Since 2005, a project run by an international group of physicists has generated a beam of neutrinos in a nearly mile-long tunnel thirty stories below the Fermilab particle physics facility, near Chicago. The tunnel points in the direction of a converted underground mine in northern Minnesota, 500 miles away, that now houses a 6,000-ton particle detector, which the neutrinos reach about 2.5 milliseconds later. Most pass through unnoticed. Of the trillions sent every year, only about 1,500 leave a mark, and these are timed via a common-view GPS system.

In 2011, a similar project that shot neutrinos from the CERN lab, on the Franco-Swiss border, to a detector in Italy's Gran Sasso

Mountain, made the startling discovery that neutrinos move faster than light. The results were later invalidated, but GPS was not the problem—it turned out that a cable connecting the GPS receiver to the master clock was loose. Remember to plug everything in correctly, and GPS time is pretty much infallible.

During the same period GPS was eliminating the need for transatlantic timekeeping flights, telecommunications technologies were increasing their dependence on precision timing. In telephony, this was especially evident in the increased use of multiplexing techniques, maximizing bandwidth by combining multiple signals into one.

In 1968, the Empress telephone exchange in London, near Earl's Court, became the first exchange to use a digital technique called pulse code modulation, with an inaugural call between the U.K.'s postmaster general and the mayor of the London borough of Hammersmith. A PCM signal allowed for a type of multiplexing that interleaves the pulses from several transmissions and disentangles them on the receiving end. That can only work if all the digital switches on the network are synchronized.

In America, the mostly analog Bell System used a type of multiplexing that kept the signals discrete, but allowed them to travel alongside one another by dividing the lines into sub-bands. To ensure that every multiplexor was on the same clock, Bell established the first national synchronization network around 1970, headquartered in an unassuming building 40 miles south of St. Louis. An ensemble of caesium clocks regulated a steady pulse, sent on its own dedicated part of the phone lines throughout the country. It was not a perfect system, but analog signals were not demanding and could handle small frequency errors.

Beginning in the late 1970s, telecom networks became increasingly digital. In the U.K., the post office developed System X, the nation's first comprehensive digital exchange, synchronized by a caesium clock, with a second clock on standby. In the U.S., the Bell monopoly was dissolved, to be replaced by several large independent companies, including long-distance carriers who saw fiber optics and other digital transmission methods as a way to maximize competitive advantage. One master clock would not provide tight enough synchronization for such a disparate network. Yet the system had to behave as one, so that the many segments of the former Bell System could interact seamlessly.

The industry players collectively agreed that the timekeeping of any primary reference source (PRS)—a clock performing synchronization duties anywhere on the network—must be sourced to an origin considered "Stratum 1," the timing community's most exacting clock standard. There were only three possibilities: caesium clocks, Loran-C, or GPS. To synchronize all the nodes of this vast system using atomic clocks was unfeasible. Loran-C was free, and the network had been in place for decades, but it did not provide nationwide coverage. GPS was free, but risky. The satellite constellation was still not complete, and steps would need to be taken to correct for the dithering effects of the Pentagon's selective availability program. But it was the best option. AT&T, the company that controlled the Bell System, retired the Missouri clock.

Around the world, telephone companies were reaching the same conclusion. New digital protocols introduced in the 1990s required even tighter synchronization, and therefore more PRS sources—numbered in the thousands for every network—nearly all of them using GPS as their time source. Today, GPS is behind every land-line telephone transmission.

For mobile phone networks, which also use multiplexing, the need for time synchronization is even more stringent. A tower needs to hand off each call signal to the next with extreme precision. Most carriers rely on GPS timing for this synchronization. A timing error of just ten microseconds can cause problems. Humphreys's team demonstrated that their spoofer could, with less than thirty minutes' transmission of its bogus signal, cause a tower to experience that deviation. They also presented evidence that the spoofer could cause multiple towers in an area to interfere with one another, and also disrupt the function that automatically identifies the location of a phone that calls 911.

Humphreys was later contacted by a rep from Alcatel-Lucent, the French network equipment company that supplies several of the world's largest mobile carriers, who informed him that the Trimble GPS receiver the company now used was spoof-proof. Humphreys says, "We promptly bought one and spoofed it."

On August 14, 2003, an electrical transmission line in northern Ohio experienced a large power surge that made it sag into a tree, causing a flashover, an electrical arc that shorted the line. That mishap caused a chain reaction throughout the Eastern Interconnection, the largest of two electrical grids that, along with a handful of much smaller regional grids, comprise the overall North American electrical grid. Within five minutes of the shorted wire in Ohio, 256 power plants spread over hundreds of miles reported failures. Blackouts left 50 million people without power, caused $6 billion in damage, and were blamed for at least eleven deaths.

Five weeks later, something very similar happened on the other side of the Atlantic. During the first hours of September 28, Italy was importing about 24 percent of its electricity from Austria, France, Slovenia, and Switzerland, all part of the Synchronous Grid of Continental Europe, the world's largest interconnection, which spans twenty-four countries. The power, about 300 megawatts greater than the normally acceptable import level, put stress on the Swiss grid. At 3:01 a.m., a major transmission line in the Swiss Alps experienced a tree flashover. Ten minutes later, the Swiss power operator in Laufenberg called his Italian counterpart in Rome and asked for an import reduction. Italy complied, but the Swiss authorities, who should have cut the power from the transmission line within fifteen minutes, let the situation fester. The power carried by the damaged line flowed into a parallel line, exceeding its capacity by 10 percent and triggering a second flashover twenty-four minutes after the first.

At that point, disaster was unavoidable. Italy was now a virtual island, disconnected from the rest of Europe. Except for the actual island of Sardinia, the entire country went dark. The blackout affected 57 million people and killed three. Ripple effects were felt throughout the grid, which experienced power fluctuations and unpredictable flows, and lines between France, Germany, and Belgium were dangerously overloaded, but the major damage was confined to Italy.

Besides the similarity of the damage and its effects, the two events had something else in common. The problems might have been avoided with better real-time information, and coordination might have facilitated better decision-making. The most common way to monitor electrical grids is a system called SCADA, which examines the power level at various nodes and sends back reports every few seconds. As much as thirty seconds can pass between SCADA observing and transmitting data—an eternity on an electrical grid, where problems beget other problems very quickly. An alternative, developed in the late 1980s by two Virginia Tech professors, was already a fixture in China and a few other countries. Phasor measurement units (PMU), also known as Synchrophasors, gained traction in the years after these blackouts, and have shown a huge surge of popularity in recent years.

A network of Synchrophasors—a "smart grid"—is like a hyperactive steroidal SCADA. Once every second, a Synchrophasor gathers 120 different types of data about the power at its node and transmits them instantly. Operators monitoring the grid can see the data displayed on a map, providing a real-time overview of power flow. Electrical grids are complex organisms that obey no national borders, with power flowing in different directions across large distances. As more power generated by alternative energy sources enters the grids, maintaining order has become even more necessary. Synchrophasor data is also useful for looking at the overall state of a grid, and noting any chronic problems.

In 2010, there were roughly 200 Synchrophasors on the North American grid—per capita, not much more than the thirty Italy installed following its blackout—mostly used for research and testing purposes. Boosted by $158 million in 2009 federal stimulus money, the number jumped to 1,700 by 2015. Around one-third of the total are part of the continent's largest Synchrophasor project, the Western Interconnection Synchrophasor Program, with ninety-seven participating power providers. Synchrophasors are popping up around the globe. Although India's Synchrophasors were in too much of a nascent state to prevent the crippling 2012 blackouts that left 700 million people without power, the country's state-owned utility company had already hailed the technology as a "revelation."

For a smart grid to have value, its Synchrophasors must observe and report at *exactly* the same moment. Because they are spread over a large area, the easiest way to synchronize them is by connecting them to high-precision clocks sourced to GPS. If someone were to introduce a bogus GPS signal that disrupted the clocks and broke the synchrony, causing an distorted and possibly alarming overall view of power flow on the grid, what might happen?

For now, probably not much. The technology is at an intermediate stage, not yet considered a "critical cyber asset," a classification used for hardware, software, data streams, and networks whose disruption could bring key parts of the critical infrastructure to the brink of disaster within fifteen minutes. But the next step is to make these smart grids smarter, giving them the ability to take direct action. They could redirect power to allocate it more efficiently or safely, and even shut the whole mess down, killing the power for thousands of users to isolate a problem before it spreads. If a spoofed GPS signal distorted Synchrophasor data, human operators might sense something was askew before taking action. Left to its own devices, the grid itself might not.

One of the few current examples of PMUs being used for actual control purposes is a line that links two hydroelectric dams on the Grijalva River in Mexico. "If somebody went down there and set up a couple of spoofers at strategic locations and spoofed the PMUs, the power transmission lines would pop," says Logan Scott, the security expert whose early writing on spoofing inspired Todd Humphreys to investigate the concept in grad school. "It's not going to be a permanent-damage kind of thing, but there's going to be an over-voltage condition. Transformers are going to take themselves offline, and, long story short, there's no electricity coming out of the wall socket."

According to accepted to international standards, if a Synchrophasor is off its timing by as little as 26.5 microseconds, its data compromises the grid's integrity. In his ongoing effort to demonstrate the toilet-tissue-thin protection accorded to GPS receivers, Humphreys tackled the Synchrophasor problem in his lab. He applied his trusty GPS spoofer to a Synchrophasor's clock and induced a 26.5-microsecond glitch.

Automation is on the horizon for the electrical grid, but it is already in full force in the financial services industry. The system cannot function without accurate timing. This is especially true now that at least half of all trading on the major exchanges is automated. The major exchanges themselves are now inseparable from the computers that do the trading. Meanwhile, high-frequency traders specialize in exploiting the latency inherent in markets, programming their computers to make automatic trades that utilize information before the market as a whole can fully react. For these traders, and the trading algorithms they build, tiny fractions of a second make all the difference.

Since around 2005, data providers have catered to high-frequency traders' insatiable appetite for speed by constructing ever-faster communications lines between the cities where most of the world's heaviest trading volume occurs. That war is nearly over. There is nary a stray microsecond left to shave off these travel times. The logical next step has been for high-frequency traders to decrease the physical distance between their computers and the exchange computers as much as possible. For a fee, the exchanges allow traders to "co-locate," housing their trading computers in the same facilities as the computers that process trades, with all the various traders' computers connected to the exchange computers by identical lengths of cable, so that even the *position in the room* cannot impart an unfair advantage. "A decade ago, a tenth of a second was an acceptable time stamp resolution," Humphreys said in his congressional testimony. "High-frequency traders now demand nanoseconds."

The potential problem with these arrangements is that the exchange clocks, like so many extremely accurate clocks in this world, receive their time signal from GPS. Humphreys, who has met with exchange officials and observed their setup, believes they have the necessary safeguards in place to make a spoofing attack on their clocks very unlikely. He does not have nearly the same level of confidence for the co-located high-frequency traders.

The exchanges give their co-locator clients two options. They can "slave" their trading computers to a time source provided by the exchange: usually an atomic clock, linked with its minions via precision time protocol, a very accurate and dependable way to synchronize internal computer networks. The use of the atomic clock—which runs wholly independent of GPS time—along with the internal safeguards the exchanges employ to detect GPS spoofing attacks, makes this the most secure choice.

Many high-frequency traders choose option two. They feed their computers the raw GPS feed coming from the antenna on the building's roof. After all, the exchange's atomic computer may offer sub-microsecond accuracy, but that necessarily means that time in its universe lines up with GPS time. Maybe *one millisecond after midnight*, for this clock, occurs *one microsecond after one millisecond after midnight*, as defined by GPS. If your trading algorithm is programmed to do something at that moment in time, it may lag behind other traders who use GPS time. The problem, Humphreys maintains, is that the traders' computers do not have the anti-spoofing protection the exchange applies to its own computers. By jacking directly into the GPS data stream, they leave themselves vulnerable to a spoofed signal that scrambles their computers' clocks.

What might the results of a successful spoofing attack on trading computers look like? Humphreys thinks it could cause a more catastrophic version of the 2010 Flash Crash, a thirty-minute hiccup when the major markets all but collapsed and then quickly rebounded. Though the cause of the crash is still debated, some evidence points to automatic trading programs used for high-frequency trading, which have instructions to pull out of the market if the program senses a problem. In his testimony, Humphreys noted that the crash revealed that many trading algorithms included automatic checks triggered by "unusual latency" in the data coming from exchanges. "In other words, if transaction time stamps don't look right, algorithmic traders flee the marketplace," he said. "The high-frequency traders who own the servers do not like inexplicable market behavior, and unlike old-fashioned traders who are obligated to stay in the market no matter its behavior, high-frequency traders can pull the plug at any moment."

A spoofing attack would have a similar effect on automatic trading as on Synchrophasors, sowing enough confusion to make them react disproportionately. The difference is that even with all the ties between interconnections, the pieces of the electrical grid remain at least somewhat self-contained. If a tree flashes over in Ohio, it doesn't make a sound in Rome. Global capitalism respects no such boundaries.

"The military *hates* Humphreys," Nunzio Gambale was saying. "He's persona non grata." Gambale's Australian accent really did a number on those vowels. "He's young and he's very ambitious," Gambale continued. "He used to be a magician, so he's got the chutzpah. And the *balls*. All those tests where he made the helicopter scream toward the ground? Those are the guys I'm talking to. The problem is, he signed an NDA with the government, and then just went out there and presented [his results]. And what can they do? It's too bloody late! He wants notoriety and he's got it."

It's true that Humphreys worked as a magician in college. The claim about nondisclosure seemed dubious in the extreme. As Humphreys points out, what kind of academic would he be if he conducted experiments without publishing the results? ("I was very clear about this with DHS.") But Gambale was in an expansive mood, and he was speaking from a position of some authority. His Canberra-based company, Locata (pronounced like an Australian saying the word "locator") had recently won a contract to install a "truth reference system" at White Sands, an ultra-accurate coordinate grid for the GPS testing activities that occur there. I was meeting Gambale for lunch in New York shortly after he returned from meeting with Homeland Security officials to pitch Locata's services as a way to mitigate the dangers of GPS spoofing.

"You should ask yourself, Greg," he said, "how does Nunzio, whose company is from Australia and has thirty-seven people, get an audience with Ron Hewitt, the head of infrastructure for Homeland Security, and a *shitload* of his guys?"

I was already asking myself this. Nobody at DHS would talk to me.

Gambale smiled. "It's because they have problems," he said. "But they're not gonna tell you that!"

In 2004, three years after the Volpe Report warned of GPS's vulnerabilities, the White House directed Homeland Security and the Department of Transportation to jointly research, develop, and maintain backup systems for GPS-dependent facets of the critical infrastructure. Eight years later, Homeland Security released an unclassified "national risk estimate," warning that "U.S. critical infrastructure sectors are increasingly at risk from a growing dependency on GPS." An investigation by the U.S. Government Accountability Office, issued in late 2013, criticized Homeland Security for its inability to "ensure that critical infrastructure sectors could sustain essential operations during GPS disruptions," and that little progress had been made since the 2004 presidential directive.

"Let's go through the problems," Gambale said. "GPS is a seventies' technology. Name one other technology put together in the era of the Beatles that is still considered cutting-edge. The Brad Parkinsons of the world designed GPS for a global military imperative. There was never an intention for it to be used for infrastructure, for timing. The technology is exquisite. I actually think it's one of the greatest inventions of the twentieth century. It's an absolute testament to Parkinson and his team that the technology has now been stretched so far beyond the design parameters. But you can't modify it at a rate that's commensurate with modern electronics. There's now a massive dependence on time. You can't guarantee availability. Spoofing is only getting better and more powerful, and it's going to be a serious problem. Parkinson calls GPS the stealth utility. That's the mother of all delusions. What's the backup?"

Locata is part of a cottage industry of technologies that provide location information in environments where GPS coverage is either nonexistent (Locata's products are popular in the mining industry) or spotty ("indoor GPS," popular in places like malls, where customers' exact locations can be leveraged by marketers). Gambale was trying to sell Homeland Security on the idea of using Locata's networks of self-regulating beacons, which do not require a satellite signal, to build localized networks—in densely populated areas, for example—to be activated in case of GPS disruptions.

A quiet consensus has formed that the best backup to the "seventies' technology" is one that predates GPS. Supported by heavyweights like Brad Parkinson and NIST timing guru Judah Levine, the idea would be to reactivate the old warhorse LORAN. The current version, called eLoran, emits a signal 1.3 million times stronger than GPS, and therefore much more difficult to overwhelm with a spoofer. As a land-based technology, one eLoran system cannot cover the entire world. But individual networks could keep the critical infrastructure from experiencing problems due to the loss of GPS. Since LORAN technology is clock-based, eLoran could conceivably perform the same clock functions as GPS.

In the U.S., the Department of Homeland Security took steps toward implementing eLoran in 2008, though funding for the program was later cut. Meanwhile, countries around the world are embracing it as a GPS backup, including India, Russia, the Netherlands, and South Korea, the victim of several pesky GPS jamming attempts by North Korea. The U.K. has taken a lead in eLoran implementation. An eLoran system now regulates shipping traffic through the Dover Strait—"the part of the English Channel that is the world's busiest choke point," David Last, former head of the Royal Institute of Navigation and an emeritus professor at the University of Bangor, told the journal *Inside GNSS*. "We get something like 500 ships a day coming through it—and ferries dashing back and forth across it. The gap they are coming through is narrow enough that you could take out the whole of the shipping activity there—take out the GPS activity—using a fairly low-cost jammer on either the British or French side."

As for what needs to be done to safeguard against spoofing, there is little agreement. Logan Scott, whose early writing on GPS vulnerabilities first inspired Humphreys, believes that although Humphreys's drone-smashing antics play well to the cameras, the potential problem he highlights—radiofrequency spoofing—may be diverting needed attention from a more prosaic type that is already occurring. Rather than trick a GPS receiver by feeding it a spoofed signal to obey, this other technique involves users controlling the transmissions coming from their own receivers: self-sabotage, in other words. The first hit a Google search for "GPS spoofer" returns is a Google Play link for a "free GPS location spoofer app" that promises to "overwrite your current location elegantly and the third [sic] party apps will think you are in Paris under the Eifel [sic] Tower or in New York on Times Square! Prank your friends on any social network to think you are somewhere else."

You *could* do that. Or you could use the spoofer for privacy purposes, to foil all those third-party apps that, for shadowy data-mining purposes, always want to "use your current location." You might also use it to convince a service provider to stream content not available in your area. It might come in handy if you live in Slovakia, where nearly 18,000 kilometers of streets and highways are tied into the world's most rapidly expanding GPS-based road toll system. (Germany, Hungary, and Switzerland are among the countries that have implemented similar plans.) The app maker anticipates these nefarious uses with the disclaimer that "any foul usage will not be supported by our team."

Though he long ago abandoned GPS to assess other vulnerabilities, Roger Johnston, whose team performed the earliest tests of the spoof threat, still believes the spoofing problem could be solved by some simple tweaks to the GPS signal that would make it easier to authenticate. Humphreys also advocates improved signal authentication, but using more advanced cryptographic techniques. Johnston believes that Humphreys both overplays and underplays the threat, exaggerating both how hard it is to spoof a GPS signal without detection and how much damage could be done with a simple satellite simulator. Johnston still thinks we need to worry about Grandma as much as James Bond.

In the spring of 2013, Humphreys gave a talk at the South by Southwest Festival in Austin. In the audience was Andrew Schofield, an amateur scientist and captain of an $80 million "super yacht" called the *White Rose of Drachs*, owned by a Briton who chose to remain anonymous. Schofield was impressed by Humphreys's work with the drone, and he now proposed that Humphreys's team attempt to spoof the *White Rose* as it traveled from Monaco to Rhodes, Greece. Humphreys and two graduate students soon decamped to Europe to see if they could trick a 213-foot yacht.

Although very little of the world's freight is ferried on yachts, 90 percent of it does travel on the oceans. And 100 percent of those ships rely on GPS—even more so than land conveyances, which, no matter how lost they become, don't face the vastness of the open ocean. "Cars aren't gonna drive off the road, trains won't ride off the rails," says Rick Hamilton, a GPS expert at the U.S. Coast Guard's Navigation Center in Virginia. "When a ship's coming into port it's got radar, visual bearings, range lights—a number of things to fall back on. But when you're outside land, there's nothing."

You can guess how this trial turned out. When the *White Rose* was cruising in international waters 30 miles off the coast of Italy, the team made its move. They perched on the upper deck and aimed the spoofer at the yacht's two antennas. As with the Hornet, they started their signal at a whisper, slowly raising it until it overwhelmed the real GPS signal. They entered coordinates that positioned the ship three degrees off course, making it turn slightly to the right. Then they let the ship's crew unwittingly compound the problem. Faulty data from the spoofer made the ship's navigation system think it was drifting left, so the crew initiated a maneuver to move it to the right to bring it back on course. Soon the ship was hundreds of meters off course.

"It's pretty breathtaking, really," Humphreys said when the results were made public, emphasizing that the data showed the experiment could work without the spoofer being on board. "You could be miles away on another ship. If you were airborne, you could be 20 to 30 miles away. All that matters is that by the time your signal arrives at the vessel, it's stronger than the real signal."

The *White Rose* hijacking was a kind of denouement for Humphreys. He'd made his point; it was time to move on. When I emailed him to ask if there was anything left to spoof, he wrote back, "I think we're done with this kind of grandstanding for now."

CHAPTER SEVEN

Better Living Through Tracking

The most dangerous two miles in America, according to the FBI, terrorism experts, local law enforcement, and government officials, is in northern New Jersey, extending from Newark Liberty International Airport to Port Elizabeth. Just across the Hudson River from the site of the World Trade Center, this stretch of mega-infrastructure has it all: refineries, major rail lines, shipping, chemical plants, oil tanks, pipelines, and one of the largest seaports on the East Coast—just about everything on the critical infrastructure checklist, excluding only dams and agriculture. Among the many possible terrorist targets in the area is a plant that processes chlorine gas—and stores 2 million pounds of it on site; the government has declared that an attack on it could threaten the lives of 12 million people, or nearly one out of every twenty-five Americans.

Look up, and you realize that this is a problem with three dimensions. Newark Liberty is one of the country's most bustling airports. For both international flights and cargo transport, it is among the top five. Every day, more than 1,100 planes pass through Newark Liberty, sharing airspace with two more of the country's twenty busiest airports, John F. Kennedy International and LaGuardia.

Slicing through this potential inferno is the New Jersey Turnpike, a segment of Interstate 95, the eastern seaboard's major connective artery. "If you want to make New Jersey the heart of America or the heart of the Northeast," the state's Homeland Security director has said, "the Turnpike is its aorta." Every day, it pumps a steady heartbeat of 100,000 cars, trucks, and buses through these two miles, Airbuses dropping from the sky on one side, airborne particulates rising on the other.

It is an area that requires extreme precision, with many moving parts passing perilously close, a constant threading and rethreading of many needles. What better place to serve as a proving ground for the next chapter in GPS-enabled commercial aviation? This would be the third installment of what Per Enge, the Stanford engineer, called the "integrity trilogy." It began with the guidance system that lowered the blood pressure of Alaska Airlines pilots as they descended into Juneau. Next came WAAS, the Wide Area Augmentation System, accurate enough to merit those seven 9s, so that any pilot could now land a plane with instruments in lieu of eyes.

WAAS was a hit among smaller carriers and private planes. For the large airlines, it wasn't as much of a revelation. Most of their planes were already landing on runways equipped with radio beacons that they could use for instrument-only approaches. This third innovation was for them. It was called the Ground-Based Augmentation System (GBAS), a setup of four highly accurate GPS receivers arrayed near one another, one GBAS setup serving an entire airport, calculating and transmitting to pilots extremely accurate GPS corrections. The primary reason the major airlines like GBAS is that the system is a requirement for some of the new approach methods now allowable with GPS.

In 2009, Newark Liberty officials expressed interest in GBAS. Working with the FAA—and with the cooperation of Continental Airlines, Newark's largest passenger carrier—the airport prepared to test the system.If the tests were successful, Newark would become the first airport in the country to use GBAS.

The tests would also help fully validate GBAS, which had not yet been cleared to handle instrument landings in conditions when visibility approaches nil. Continental planes would have GBAS receivers in the cockpit, though the pilots would still use the airport's runway beacons for landings. The GBAS data would be collected and analyzed to determine how the system had performed. In late November, the system was ready and testing began.

The system faltered almost immediately. Several times a week, it would take itself offline and not come back to life until hours later. The interruptions continued through the first weeks of 2010. Something was affecting the ability of the GBAS setup to receive its GPS signals and send the corrections on to the approaching planes. At first, the testing crew was stumped. There were apparently no problems with the system itself, no signal problems or software errors—but something was consistently poking holes in GBAS's secure space.

By March, they had determined the source. It was flowing from America's aorta. Someone on the Turnpike was stamping out the signal.

Ralph and Robert Schwitzgebel were identical twins from Ohio, champion high school debaters who won the state title in 1951, graduated from different colleges, and both—unbeknownst to the other—applied to Harvard's graduate program in psychology. "We kind of show up on campus one day—'What are you doing here?' " Robert recalls.

It was a heady time at the Harvard psych department. The faculty included B. F. Skinner, behaviorism's leading figure, and also Timothy Leary, who demonstrated during his brief time at the university that he was willing to go to unprecedented lengths to test the molding of human behavior. Leary became Ralph's adviser. Ralph coauthored the paper detailing Leary's infamous

Concord Prison experiment, in which young inmates were given psilocybin as part of group therapy, between 1961 and 1963. The study proposed that the drug had a positive effect on the recidivism rate of the experimental group.

Ralph took from his mentor a willingness—even an eagerness—to deploy unorthodox methodologies, especially in the treatment of young people on the margins of society. Ralph wanted to merge the experimental psychologist's lab with the psychotherapist's office. In 1959, he founded the Science Committee on Psychological Experimentation (SCOPE), using grant money to counsel gang members and other at-risk youth in the Cambridge, Massachusetts, area. He believed that the therapy methods traditionally used with juvenile delinquents seldom worked because of a culture clash between "delinquent" and "nondelinquent" cultures. Accordingly, he did not consider these people his patients so much as his employees. He would approach them on street corners and offer them money and equipment to film their lives, keep audio diaries, and submit to interviews. Over time, he would win their trust, and their behavior would begin to change. It was a form of stealth therapy.

SCOPE's work was controversial. Some psychologists criticized Ralph for advocating a "soft" approach that coddled the kids, failed to address the root causes of their delinquency, and had no long-term positive effect on their personalities. Ralph countered that he was taking a practical approach. He wanted to affect behavior in a way that reduced the likelihood that the kids would commit crimes and get themselves into trouble right now. Considered decades later, much of SCOPE's work seems ahead of its time—especially the emphasis on employing recording media for self-expression—while some of it now seems glaringly retrograde. In a *New York Times* profile of Ralph five years after SCOPE was founded, he cited, as evidence of the efficacy of his behavior modification methods, a youth who came to SCOPE for help with "homosexual tendencies." His treatment regimen involved drinking ipecac when the urges grew strong, and he was now married and "very content," Ralph said.

In *Streetcorner Research*, a book Ralph published in 1964 that detailed his work, he asserted that the "nondelinquent" culture seldom interfaced with the "delinquent" one in a way that contained enough "intensity" to break through and elicit behavioral change. An alternative approach would be to develop "a humane technology which will eliminate unwanted behaviors and develop in their place desirable behaviors." He was talking about positive reinforcement, a key tenet of Skinner's approach to behaviorism. Skinner was Robert's advisor. Robert was on board with SCOPE's project, and became his brother's chief collaborator.

The Schwitzgebels disdained pigeons, Skinner's experimental animal of choice. "Pigeon data was really boring, but the reinforcement idea seemed really powerful," Robert says. "Certainly juvenile delinquents and gang members were a lot more interesting than pigeons, so let's just go out and get some, hook them up, and use reinforcement. They don't cost much more than pigeons, they're more interesting, and they can be very cooperative. So that's what we did."

The freeform atmosphere in the department favored this kind of convention-flouting. "No forms, no institutional review committees—we didn't have to do any of that," Robert says, laughing. "This was the days of the sixties and Leary."

Ralph had an epiphany while seeing the movie *West Side Story*. What if there had been some communications system that could detect when the gang was about to fight, putting them in touch with people who could talk them down? Ralph envisioned a small transmitter, worn on the body, that could detect certain forms of behavior and transmit data to a base station for analysis. The station could send back signals in a sort of "behavioral feedback system that may have considerable therapeutic potential." Although that idea never came to fruition, the brothers managed to construct a different kind of monitoring system.

At a cocktail party, Ralph met an engineer, and the two got to talking about Ralph's *West Side Story* idea. The engineer was intrigued. He happened to have a line on some surplus military missile tracking equipment, which he helped the brothers set up in a former car showroom they rented. With this, they built the world's first functioning electronic monitoring system. Over an area of a few square blocks, the brothers erected transponders, on places like the tops of telephone poles and behind the cross on the roof of the Cambridge Baptist Church. The beacons detected the presence of anyone wearing a receiver contained in a two-pound box, worn on a leather belt. The person's location, updated every thirty seconds, would show up as a light on an electronic wall map at the office.

To recruit subjects, they'd strike up conversations with rough-looking teens on Cambridge street corners, saying they were doing research on what kids thought of police. Those who seemed interested were invited back to the Schwitzgebels' office the next day to record some of their thoughts. The positive reinforcement would start immediately, before the kids were fully informed about the project. A kid who showed up early might receive a $15 bonus. Throwing away an empty soda can at the office might elicit praise from the brothers, who would inform the kid that because of his spontaneous good behavior he was receiving two free pizzas. The awards appeared to the subjects as random—they never knew when one was coming or what it might be, a central tenet of Skinnerian positive reinforcement. "It absolutely has to be a variable interval, so you don't know when it will happen, and a variable ratio, so you don't know how much work you have to put in," Robert explains. "It's like a junior Las Vegas. If you knew how much you would win or lose, you might not go to Vegas."

Once a subject had proven to be dependable and fit the criteria—a troublemaker, someone clearly headed down the wrong path, but without a criminal record—the brothers would let them in on the full project. A kid would be asked if he would be willing to wear a leather belt attached to a two-pound box containing a receiver, as well as an antenna and rechargeable batteries.

The point of the experiment was to watch over kids who had already proven that they could not—or would not—obey society's strictures. It would be more like a way of knowing if these youth were where they were supposed to be at any given time. If they were at school, or work, or home, or even church, that meant they were not only living their lives according to the rules, but also *not* placing themselves in areas where trouble was likely to find them. If the system showed that the kid was where he was supposed to be, he might receive a reward, but he would never be penalized for being in the wrong place. "Our idea was no punishment," Robert says. "That was Skinner's rule—no punishment, just reinforce them when they were at school or drug treatment or whatever."

From the beginning, the brothers thought of their experiment in grand terms, imagining an enhanced version of their system for parolees. Maybe the belt could somehow record physical and physiological variables, such as nervous gestures and anxiety, or even an increased blood alcohol level, sending an alert if data suggested that the wearer was in a state to do something rash. If the crime involved the use of a motorcycle, perhaps the device could measure the speed the wearer was traveling. Tones could be used as codes. Monitored people who showed up at work might receive a "good job" tone; staying out late might elicit a "return home" tone. The Schwitzgebels never saw their approach as purely a matter of surveillance, but rather an efficient and effective method of positive reinforcement that would reduce recidivism, since parolees would know there was always a record of their behavior. They envisioned the day when prisons were obsolete, and became "museums or monuments to the inhumanity and ineffectiveness of social retribution."

Although these plans never solidified, the Schwitzgebels continued to run their program for wayward youth, and were pleased with the results. They established a small nonprofit with money they received for selling the rights to their story to Universal Studios. (A movie was never made.) The program finally folded in 1975. The brothers had failed to persuade many people that electronic surveillance and positive reinforcement were a winning criminal justice combination. To critics on the right, "positive reinforcement" smacked of coddling criminals. To those on the left, it seemed ludicrously Orwellian. In *Kind and Usual Punishment*, Jessica Mitford's 1973 book about the problems with the American prison system, she took Ralph to task for his advocacy of electronic monitoring: "Is there also a broad hint for the rest of us concealed in here somewhere? For if the 'behaviors in the community' can be electronically spotted and corrected for parolees, why not for the entire population?"

"My brother got a really hostile letter from the editor of *Probation* magazine, saying this is really Big Brother stuff," Robert says. "It said the next thing that would happen was it would be implanted, and the devices would be 'Big Mother' or something of that sort." The brothers both moved into academia. With their ideas of electronic monitoring seeming fanciful or fascist, depending on the critic, the Schwitzgebels' dream of better living through tracking appeared to be over. It would, in fact, be resurrected, in a form the brothers hardly recognized.

At the same time the Schwitzgebels were laying the conceptual and (to a lesser degree) technological groundwork for electronic monitoring of offenders, the criminal justice system was discovering the legal ramifications. In the U.S., this involved debates over the Fourth Amendment's guarantee against illegal searches and seizures—and how this edict dovetailed with the concept of privacy in the context of new technologies.

The first important test case, *Katz v. United States*, reached the U.S. Supreme Court in 1967. FBI agents had placed an electronic listening device in a phone booth they knew their target used to call in illegal gambling wagers. The court ruled that this practice violated the man's Fourth Amendment rights, because although the phone booth was public space, its usage implied a reasonable expectation of privacy. The decision established the idea that Fourth Amendment protections did not begin and end with one's home or private property, but also extended to spaces one occupies temporarily.

Throughout the 1970s, police departments began to deploy pre-GPS surveillance setups called beepers, which used a receiving antenna and a radio frequency detector to follow the signal emitted by a surreptitiously placed transmitter. Police would manually adjust the antenna in search of the signal, like pre-cable television viewers manipulating the rabbit-ear antenna attached to their sets. Sometimes police in a car would hang an antenna off of each side to see which one got the stronger signal. The cleanness of the signal pointed to the suspect's direction, while a signal-strength meter gave an indication of distance. Some officers claimed their ears were better than the meters and "trained" themselves to hear the difference. The range was as small as two blocks in a congested area, two to four miles on an open road, and about 20 miles if the beeper was in a helicopter or otherwise airborne. As there was no map component, the beeper setup amounted to a crude form of dead reckoning, imparting a rough sense of where the transmitter was in relation to the receiver's location.

Using a beeper to track a suspect was the conceptual converse of putting a bugging device in a phone booth. The latter was all about enhancing observation of a single space, waiting for a suspect to move into it. A beeper involved observing a device in the possession of a person, tracking the suspect's mobility.

In 1983, the Supreme Court heard *United States v. Knotts*, a case that turned on whether this was a meaningful distinction. Employees of 3M in St. Paul, Minnesota, suspected that a former coworker had stolen chemicals used to produce methamphetamine. Based on their tip, police trailed the man and witnessed him buying similar chemicals from another company. That company agreed to let police install a beeper in a container of chloroform to be sold to the suspect. The sale went as planned, the man loaded the materials into his car and drove away, with police following the beeper signal to the home of another man, who received the chloroform and transferred it to his car. Police continued to follow this car as it crossed the state line into Wisconsin. Then they lost the signal.

An hour later, a police helicopter picked it up again. The signal suggested that the package was no longer in motion, and it was traced to a rural cabin. Observing the cabin, the police saw the suspect move the chemicals outside of it. Swiftly obtaining a warrant, they searched the cabin, discovering extensive lab equipment and 14 pounds of pure amphetamine. They made arrests, and won convictions that broke up a drug ring.

The police had fulfilled their Fourth Amendment obligations (or so they assumed) by getting a warrant to search the premises. After all, their surveillance had yielded a plethora of probable cause. But what about the tool that had furthered this surveillance—the beeper? A lawyer for one of the defendants reasoned that using the beeper was itself a search, requiring a separate warrant. The argument didn't fly. The Supreme Court unanimously upheld the conviction, on the grounds that the privacy one could reasonably expect inside a domicile—the cabin—did not extend to the public roads that led the police to it. Following someone on that public space did not constitute a search, even if the beeper made that search easier and more comprehensive. "Nothing in the Fourth Amendment prohibited the police from augmenting their sensory faculties with such enhancement as science and technology offered them in this case," Justice William Rehnquist wrote.

Knotts was a line in the sand—or, more accurately, the distinct absence of one, washed away by the high tide of technological progress. Regardless of how greatly technology magnifies our senses, noting the movement of a person through the world did not amount to a search of that person, or that person's possessions. When you were in public, you were subject to the public's gaze.

A year later, the Supreme Court decided a similar case, *United States v. Karo*, and decided that there really was a line. With his consent, federal agents installed a beeper in a can of ether belonging to a government informant, who then sold the ether to people who used it to extract smuggled cocaine. Although the court rejected the idea that the tracking itself required a warrant—since the can belonged to the agents when they installed the beeper, there was no privacy violation—the justices had a problem with how the tracking had proceeded.

The *Knotts* police had let the beeper lead them to the cabin, but they glimpsed the movement of the telltale chemicals outside the cabin with their naked eyes. The federal agents in *Karo* allowed the beeper to do more of the work for them. Rather than staying on the suspect's trail and then seeing what they could see, they simply traced the beeper's signal to a house in Taos, New Mexico. From this, they could confidently assume that the ether was inside. Technology had given them rudimentary x-ray vision that allowed them to look through the walls of the house to "see" the illegal activity inside. Although monitoring a beeper is "less intrusive than a full-scale search," Justice Byron White wrote, this beeper had revealed "a critical fact about the interior of the premises that the Government is extremely interested in knowing and that it could not otherwise have obtained without a warrant." The use of the beeper in this case, the court decided, constituted an unlawful search.

It was just a few years since the launch of the first GPS satellite. As the constellation grew—and, along with it, a method of tracking that further enhanced the ability of humans to observe and gather data—the question of sensory augmentation and the law would grow more urgent.

For an airport of its size, Newark Liberty is squeezed into a fairly small footprint. The team that set up the GBAS receivers found that the only spot at the airport that provided full coverage was next to a runway at the airport's eastern edge. Each receiver sat within 200 meters of the New Jersey Turnpike.

The freeway's centrality guarantees that a sizable amount of its traffic is always commercial vehicles. A significant number of these, it turned out, are driven by people who possess "personal privacy devices," a euphemistic term for GPS jammers. The drivers carried these jammers to foil the tracking apparatuses set up by their bosses to monitor the whereabouts of employees. Although operating one is illegal, they are cheap—often under $100—and easy to obtain from Internet vendors. Small and innocuous, many plug right into a vehicle's dashboard cigarette lighter. They form a little interference bubble in the area surrounding the jammer, stopping the GPS signal from getting through. This interference was disrupting the GBAS receivers.

The discovery came as a shock to the GBAS team. "We were generally aware of people with jammers in the cars," says Sam Pullen, leader of the Stanford GPS Lab's GBAS research group. "But we knew those devices were low-power. So we suspected there would be fewer bumps and that we'd rarely see them. We didn't think about it that closely, but there was no reason to think that car-based jammers by themselves would be much more frequent than what we had seen previously."

Authorities began a stakeout of the Turnpike. But pinpointing the source of the jammers was difficult, so the GBAS outages continued. On April 29, the effort finally paid off. Police were positioned on the Turnpike, near the runway, when the interference began. It seemed to coincide with the passing of a certain truck. They raced after it and motioned the driver to pull over. The jammer was right there on the dashboard. The driver made no attempt to hide it. In exchange for handing it over, the police let the driver go on his way with a warning.

Moving the GBAS array was not an option. No other spot could so effectively cover the airport. Instead, the team made some adjustments to the software, which mitigated the problem a bit, but not completely. Cars and trucks with GPS jammers continued to use the Turnpike, and sometimes they jammed the GBAS. Throughout the summer months, the team monitored the problem—the monotonous jamming blips from trucks passing through America's most dangerous two miles.

Until one day in August, when the static suddenly got worse.

By the second half of the 1980s, companies like Trimble and Qualcomm were exploring the market for GPS devices that would allow trucking companies to monitor their drivers' whereabouts. A GPS tracker is simply a GPS receiver integrated into some kind of communication device that periodically transmits or records that location. The problem for early fleet management systems was not so much the GPS aspect, as it was finding radio frequencies usable over wide geographic areas to transmit the coordinates back to the monitoring centers. But the customers were there. Three years after it debuted, Qualcomm's Omnitracs system had signed up about 100 trucking companies and was monitoring the whereabouts of 15,000 trucks. Drivers were required to enter messages onto a dashboard-mounted terminal. Satellite dishes on the trucks relayed the information to Qualcomm's San Diego headquarters, which forwarded it to the driver's dispatchers.

As cellular networks developed and the cost of access decreased, GPS tracking became a more realistic proposition. Between 1992 and 1997, U.S. Census Bureau "vehicle inventory and use" surveys found that the percentage of commercial trucks on the road being tracked electronically rose from roughly one in ten to nearly one in four. The process of tracking via GPS became much simpler. "A GPS tracking device is like a mini-cell phone, more or less," says Ryan Driscoll, the marketing manager at GPS Insight, an Arizona-based company that designs GPS tracking software. At regular intervals, the device transmits the GPS reading to a monitor, either a live human or a computer that gathers and archives the location data.

By 2005, companies and government organizations were using GPS to track 1.3 million fleet vehicles. Analysts projected that the North American fleet management market alone would grow to $7 billion over the next few years. More than half of all fleets with 100 or more vehicles now use a GPS fleet management system—for companies with more than 350 vehicles, the adoption rate is approaching 60 percent. The worldwide fleet management industry, valued at $12 billion in 2014, is on track to be worth more than $35 billion by 2019, with Asian and Pacific markets showing the highest growth rates. In Delhi, a public-private partnership between the Indian government and an infrastructure investment company has installed GPS trackers on all 60,000 public rickshaws, to prevent drivers from gouging tourists by taking longer routes. In China, GPS tracking is used to control the black market in "gutter oil," recycled cooking oil that is combined with industrial waste and other effluents, which may account for 10 percent of all cooking oil used in the country. GPS trackers on garbage trucks keep track of where oil is collected, so that officials can ensure that the amount collected does not mysteriously decrease before disposal.

Fleet management is one of the fastest-growing segments of the overall GPS industry, and the largest segment of the telematics industry—companies that specialize in processing and transmitting real-time data from vehicles. It has proven itself to be recession-proof. "Our businesses thrive in a bad economy," Driscoll says. "When expenses are tight, they need this. They can't afford to waste fuel. They can't afford to pay drivers for wasted labor hours. They've got to shave as many costs around the entire operation as possible."

Companies that purchase fleet management services typically report that they recoup the costs in about a year. When employees who drive company cars know their bosses are watching them, they are more likely to take the shortest, quickest routes and to avoid side trips for personal business—especially when fleet management systems are augmented with larger telematics packages that can generate data such as "idle alerts" for vehicles that remain stationary for a certain amount of time. "As soon as the employees knew we were tracking, they stopped idling," a rep from the Delaware Office of Management and Budget reported, leading to an immediate 5–8 percent drop in fuel consumption among the state government's 2,100 vehicles. The water department of New Haven, Connecticut, claimed that "being able to control excess idling" saved it $64,000 in annual fuel costs among its 300 vehicles. Gold Medal Bakery, a Massachusetts-based bread products distributor with a fleet of 120 trucks, pegged its annual savings at around $167,000 in "fuel and mileage savings" and $127,000 in "driver time savings."

Among the private trucking fleets that use GPS tracking, two-thirds integrate the system with other data gathering, to analyze job performance. The most extreme example may be United Parcel Service, whose drivers—"among the most regimented workers in the country outside those on an assembly line," according to labor journalist Jane Slaughter—use trucks outfitted with 200 telematics sensors, recording such metrics as how hard a driver brakes and how long it takes to deliver a package and get a signature. In much the same way that Todd Humphreys described

GPS in a drone as the "bulwark," surrounded by and supporting other navigation tools, GPS location tracking in fleets is the foundation supporting many kinds of telematics data. Perhaps a clue as to why GPS tracking is so seductive as a gateway to other information is that fleet managers overwhelmingly rate "vehicle location" as the principal benefit of GPS tracking, far ahead of fuel consumption and cost control. As the sales manager of a GPS fleet management company put it, summing up the prevailing attitude of his clients, "I want to know my employees are working, not at the Circle K having a soda."

There is something appealing about effortless surveilling, and GPS scratches that itch. It's nice to save hundreds of thousands of dollars and conserve gas, but perhaps when we peel back the layers of efficiency concerns, GPS provides the possibility of omniscience, unlike any previous technology. There is, after all, nothing "natural" about using GPS to keep a continuous inventory of the world's moving parts. It reflects a choice, a conscious application of a neutral technology—just as Brad Parkinson's team saw a way to keep pilots safe and pinpoint their bombs. GPS tracking, which could not exist without a separate communications infrastructure, is another reminder that GPS itself is a blank slate onto which we project our desires. And what we desire most from it is perfect knowledge of other people's location and behavior.

What makes this such a combustible issue is that, as the history of law enforcement surveillance demonstrates, GPS is a powerful enough technology to take surveillance into uncharted legal waters.

In 1993, German police began investigating a man they suspected of carrying out bombings as a member of the Anti-Imperialist Cell, a two-man offshoot of the far-left Red Army Faction. They trained video cameras on the entrances to his flat, intercepted calls to his home and a nearby pay phone, and opened his mail. Two years later, they installed two beepers on a car belonging to a suspected accomplice, which the two men discovered and destroyed. Police then installed a GPS device in the car, which tracked and recorded the car's movement once every minute for the next three months. Police used this GPS data to build a successful case against him.

The man appealed his conviction, on the grounds that although German law permitted police to use "technological means" to conduct visual surveillance, that term by itself was too vague to apply to GPS tracking. Because GPS, unlike beepers, enabled effortless twenty-four-hour surveillance, police could easily construct a detailed timeline of his movements. In 2005, Germany's Supreme Court upheld the use of GPS, reasoning that although GPS surveillance required less effort than beepers, any type of electronic monitoring of movement is still less invasive than eavesdropping. The court also rejected the argument that the GPS tracking—by helping the police direct its use of other surveillance methods, such as wiretapping and mail interception—allowed the government to learn too much about this man's life. The court did, however, affirm that excessive surveillance was a potential problem that required interagency cooperation. (Germany later amended its laws so that GPS tracking requires a court order after one month.)

That reasoning underlined a subtle difference between German and American approaches to privacy. The U.S. Supreme Court, in its *Knotts* decision, declared that a person driving on public roads "has no reasonable expectation of privacy in his movements from one place to another." The German court was not making this public/private distinction. Rather, it was affirming the idea that public exposure does not lead to a legal forfeit of privacy, that individuals have a right to "informational self-determination," a control of personal data against intrusive government efforts to catalog and store it. Even in upholding the use of GPS in this case, the German decision made clear that a tracking device, combined with other surveillance methods, could conceivably violate a suspect's constitutionally guaranteed rights.

The suspect had also argued that the police's use of GPS violated the European Convention on Human Rights, a treaty that guarantees everyone "the right to respect for [their] private and family life, [their] home, and [their] correspondence"—violable only by lawful state actions that are "necessary in a democratic society" to protect itself or maintain the health, safety, or rights of other individuals. In 2010, the European Court of Human Rights, which adjudicates disputes arising from the treaty, heard the case and affirmed the right to "a zone of interaction of a person with others, even in a public context, which may fall within the scope of 'private life.'" The court agreed that GPS tracking did interfere with the man's private life—by helping the police establish patterns and expand their investigation—but that it was justifiable under the "necessary in a democratic society" exemption.

But the European court, like the German court, also stated that GPS tracking "differed from other methods of visual or acoustical surveillance," in that it was less invasive. So, on the one hand, the court ruled that GPS tracking had enabled a *potential* privacy violation. On the other, it placed GPS, like a beeper, in a different category from a wiretap.

For the police in the nearly fifty countries that have signed the human rights treaty, this distinction (or lack thereof) has important ramifications. In the U.K., the law distinguishes between invasive surveillance, such as a hidden camera or listening device, and directed surveillance, monitoring of a public space. Most invasive surveillance requires judicial approval and authorization from the Home Secretary; entering private premises necessitates an additional layer of official approval. Directed surveillance, which includes any device used to track a vehicle, requires no legal authorization at all. Ireland's surveillance laws are similar. Police in the U.K. and Ireland therefore require no court order to track somebody's car with GPS. Moreover, given the European Court's reluctance to consider GPS equivalent to other surveillance methods, it is unlikely that the car's owner could successfully claim a human rights violation. For the moment, GPS tracking of suspects in the U.K. and Ireland is all but unregulated.

In the U.S., the legal debate over GPS tracking of suspects has followed a similar trajectory, centered around the case of Antoine Jones, a Washington, DC, nightclub owner targeted in a narcotics trafficking investigation. In 2004, police obtained a federal warrant to place a GPS tracker on a Jeep Grand Cherokee belonging to Jones's wife, which Jones often drove. The warrant authorized ten days of surveillance, provided that police installed the tracker in Washington. Instead, they waited eleven days, and attached it to the car when it was parked in Maryland.

When Jones's case went to trial, his lawyers pounced on the search warrant issue, moving to suppress any and all evidence gathered as a result of the GPS tracker. Implicit in this demand was the idea that the GPS tracking of Jones fit the Fourth Amendment definition of a search. Not all legally defined searches require a warrant—nobody asks for one when their bags are searched at the airport—but most do. The prosecuting lawyers did not deny that police violated the terms of the warrant. They argued that the warrant was unnecessary from the beginning.

Jones was found guilty and sentenced to life in prison. In 2012, the Supreme Court agreed to hear the case. Though widely perceived as the one that would settle the question of whether police in the U.S. need a warrant to track suspects with GPS, the actual question the court sought to answer was much narrower: was the GPS tracking of Jones a constitutionally protected search?

The government's case treated GPS as merely the latest version of the beeper. The Supreme Court had established that the use of a beeper to track movement within a private space required a warrant (*Karo*), but tracking movement on public roads did not

(*Knotts*). The police, government lawyers argued, had only used GPS to locate Jones's car on the street, which put them on the correct side of the *Karo*/*Knotts* divide.

The defense argued that *Knotts* and *Karo*, or any other beeper case, did not apply here. GPS tracking was something different, because it basically augmented the sensory capabilities of the person doing the tracking. Unlike a beeper, a GPS tracker allowed police to gather reams of information about a subject with almost no effort. Data gathered by a beeper is subjective and contingent; GPS tracking, which yields Cartesian coordinates, is objective and absolute. When an officer testified in court in a case involving beeper surveillance, the tangible evidence was limited to the officer's recollections. With GPS tracking, the data *is* the evidence, and can be entered into court proceedings. In a similar way that following GPS directions in a car involves an illusionary melding of the real and mapped worlds, GPS tracking, in the context of twenty-four-hour surveillance, involves a personal transformation, a melding of the self with the tracker. The central concept of surveillance—using human senses to gather and process information—is abandoned. GPS tracking in effect turns the human tracker into a cyborg.

The court ruled unanimously to overturn Jones's conviction, but for confusing and even contradictory reasons. The author of the main decision, Antonin Scalia, sidestepped the issue of privacy altogether, applying a literal reading of the Fourth Amendment. The police should have obtained a warrant, because they had briefly occupied private property, the Jeep, to install the tracker—ergo, it was a search. A concurring opinion, authored by Samuel Alito, openly mocked Scalia's analysis as "highly artificial," wondering what eighteenth-century situation could possibly be analogous to GPS tracking—perhaps "a case in which a constable secreted himself somewhere in a coach and remained there for a period of time in order to monitor the movement of the coach's owner." But Alito's opinion stopped short of claiming that GPS tracking was, by its nature, a search, making his interpretation more equivocal than Scalia's.

Contrary to many media reports, the case did not settle the issue of whether GPS tracking always requires a search. At most, it suggested that police, to be on the safe side, should probably obtain a warrant before using a GPS tracker. "In effect," Lyle Denniston wrote on the legal forum SCOTUSblog, "the Court seemed to have launched years of new lawsuits to sort it all out."

In 2013, a federal appeals court deciding a case similar to *Jones* ruled that GPS tracking does require a warrant, so perhaps the Supreme Court will agree to take another swing at the issue. Meanwhile, the related issue of tracking someone via triangulated mobile phone signals—and consulting the data, which mobile companies temporarily save, to reconstruct movement—remains even more of a legal gray area.

The Supreme Court's most incisive interpretation of the *Jones* case was a solo opinion authored by Justice Sonia Sotomayor. While signing Scalia's opinion, presumably because it went further than Alito's in labeling GPS tracking a search, Sotomayor saw a growing gap between societal transparency and individual privacy expectations, which laws would eventually need to address. What made GPS tracking a challenge to fit into the nation's legal framework was that the practice itself seemed consistent with notions of privacy, since it only involved public space, but the results, perhaps, did not. "I would ask," she wrote, "whether people reasonably expect their movements will be recorded and aggregated in a manner that enables the Government to ascertain, more or less at will, their political and religious beliefs, sexual habits, and so on."

The larger problem, Sotomayor reasoned, was that we live so much of our lives in public. "More fundamentally, it may be necessary to reconsider the premise that an individual has no reasonable expectation of privacy in information voluntarily disclosed to third parties," she wrote. "This approach is ill suited to the digital age, in which people reveal a great deal of information about themselves to third parties in the course of carrying out mundane tasks." Most of us will not be tracked by GPS, but we all leave digital breadcrumbs behind us. The phone company knows to whom we talk and text, the Internet provider knows whom we email, and we purchase so many goods online that retailers know everything we buy and are likely to buy. Sotomayor thought it unlikely that people would accept the right of the government to obtain, without a warrant, a list of every website they'd visited in the past year. And this is all information we volunteer without complaint. "But whatever the societal expectations," she continued, "they can attain constitutionally protected status only if our Fourth Amendment jurisprudence ceases to treat secrecy as a prerequisite for privacy."

"Stickiness" is a term online marketers apply to websites that encourage repeat visits. But as Mary Shacklett, the president of the marketing research firm Transworld Data, pointed out in the IT journal *TechRepublic*, it could also describe how GPS lets us "build situational contexts around things and people to create new meanings, associations, and 'stickiness' of disparate data." The simplest example is when we use a program like Google Maps to learn about our location at that moment, a sticky query that draws in satellite mapping, ground-level photography, and information about nearby businesses, all keyed to a GPS-derived position. Stickiness, she notes, also supports the telematics industry, as when sensors on a delivery truck in a remote area allow a company to monitor performance in difficult terrain.

GPS-enabled stickiness can yield fascinating insights about the way people and animals move through the world. Civil engineers at the University of Illinois analyzed GPS data from 700 million New York City taxi rides to learn more about traffic patterns in the city and how natural disasters like Hurricane Sandy affect them. "There is a heartbeat pattern to the city every single day," one of the researchers explained. "The data shows us the typical heartbeat, and then we look for the arrhythmia."

Stickiness gets insidious when we expand our definition of GPS to include other position indicators. Every few seconds, your mobile phone pings the nearest tower to let it know where you are if a call comes in. The mobile carriers hold this data for an indeterminate amount of time, using it to research usage patterns and tower placement. They have also discovered that in our age of analytics, this anonymous data is a lucrative source of income when sold to outside businesses. The Madrid-based global phone and broadband provider Telefónica aggressively markets its Big Data External Monetisation Model. Verizon altered its privacy policy so it could spin off Precision Market Insights, a division dedicated to selling big blocks of data relating to its mobile phone customers' movements. In one case study involving Super Bowl XLVII in 2013, in which the Baltimore Ravens defeated the San Francisco 49ers in New Orleans, Verizon determined that the ratio of Baltimore attendees to those from the Bay Area was three to one.

The "indoor GPS" field is animated by the idea that even if GPS can reach an indoor environment, there is a powerful economic incentive to know not just generally where people are, but exactly where they are. Companies can market to consumers in a crowded mall more easily if they can tell that someone is on the third floor at H & M, rather than at the Cinnabon two floors below, at the very same latitude and longitude. What better time to push a coupon for cornflakes to your phone than when Safeway or Tesco (or Kellogg's) knows you're in the cereal aisle, not the dairy case? "What companies like Google and Apple and all the advertisers care about now is the big shift from desktops to mobile phones," says Ganesh Pattabiraman, cofounder of the indoor positioning company NextNav. "Because if they can figure out where the user is, and what he's doing, they can serve up a contextual ad. If the location can be tightened to within 20 meters, or 15 meters, then it becomes much more relevant and contextual."

The intent of data collection is not always mercenary. In 2011, two researchers revealed a curious security flaw in the iPhone. Whenever an iPhone owner backed up the phone's data on a personal computer, it left behind a file that contained information about where the phone had traveled. The engineers wrote software that converted this information into latitude–longitude coordinates. The ensuing uproar prompted Apple to explain this apparent invasion of privacy. What many iPhone owners did not realize—although Apple had already revealed it previously, to little notice—was that their phone were constantly collecting data regarding the location of cell towers and Wi-Fi hotspots. This information is transmitted to Apple, which keeps a database on these locations, and uses the data to refine the iPhone's locational ability.

Complicating the legal issues around tracking is a possible cultural shift, a fetishizing of technology concurrent with an increase in personal details we are willing to share. Sam Liang led the team that created Google Maps. Experimenting with the 700 Wi-Fi routers Google had deployed near its headquarters, Liang invented the blue dot that signifies "you are here." In 2012, he launched a company called Alohar, to develop and market an app called PlaceMe that keeps an automatic log of everywhere you go, a stickiness of constantly accruing personal information. (Sometimes too sticky: whenever I dropped off or picked up my daughter at day care, PlaceMe placed me at the liquor store next door.)

To conserve battery power, PlaceMe does not do the location calculations on your phone. It sends the raw data back to Alohar's servers, which crunch the numbers and relay back the coordinates. This data remains on Alohar's servers indefinitely, although Liang says users who want their data deleted can opt out. The information is not shared with any third party, and is retained by Alohar for analytics and debugging purposes, and to improve the technology's function.

Liang is careful to avoid using "tracking" to describe what Alohar technology does—it's just collecting data, that stickiest of substances. Nobody is following you, in other words, watching you go about your business. (The data is, however, subject to subpoenas.) But "follow" has different connotations in the early twenty-first century. Liang is willing to bet that the sharing rampant on social media allays outmoded privacy concerns. "Teenagers talk on Facebook all the time," he says. "They tweet about everything they do. Their privacy level is really low. I think it will be best to target those people first, because they can accept a new concept."

Liang points to the ease with which Gmail users now entrust Google with their personal correspondence. In the early 2000s, "if you told someone how much information about yourself you would [someday] put on the Internet, they'd laugh at you," he says. "As society moves forward, I see people becoming more and more open. You'll still have your own secrets and privacy, but the more open people are, it's gonna be a better world."

After the Schwitzgebels shut down their electronic monitoring activities, the concept languished until 1977, the year Jack Love, a judge in New Mexico, was inspired by a superhero. Love was not much of a comic book fan, but he happened to see an issue of *The Amazing Spider-Man* in which the villain, Kingpin, fits the web-slinger with a special bracelet—an "electronic radar device," Kingpin explains, "which will allow me to zero in on your location whenever I wish!"

Judge Love liked the idea of tracking—as a means of surveillance and an alternative to incarceration. "He was exercised about two things," says Mike Nellis, a criminal justice professor at the University of Strathclyde in Glasgow, Scotland. "When people came out of the New Mexico penitentiary on temporary leave, he wanted a bit more confidence about how they were supervised. He also had this thing about sentencing young people to the penitentiary in the first place. You've done probation, you've done community service—is there anything else you could do?"

Love mentioned the idea to Michael Goss, a friend who worked for Honeywell as a computer salesman. Goss researched the idea, found some of the Schwitzgebels' patents, and realized he could do something similar. He made five devices, which Love started using when sentencing young offenders, but it wasn't long before his superiors ended Love's experiment. "It was a maverick operation that Judge Love was doing," said Nellis, "but he started a trend and the world has never stopped."

Goss went on to form his own company, NIMCOS, in 1982, marketing a four-ounce ankle bracelet. It put out radio signals that were picked up by a receiver connected to a phone jack, and relayed to a central computer. If the wearer moved more than 150 feet from the receiver, the signal would not reach it, and the system would send an alert. NIMCOS soon folded, but Goss went to work for BI Inc., the first company in America that saw a large market for electronic monitoring. The first wave of these house arrest systems—setups involving radio transmitters and modems—arrived in 1987. That year, about a thousand were used in the U.S., and by 1994 the number shot up to 67,000.

Florida was the first state to fully embrace electronic monitoring. In the late 1980s, Richard Nimer was hired to run the Florida probation and parole office. At the time, the office was using early systems based on radio and voice verification. "I had a couple of major incidents where people left the house when they weren't supposed to, the computer showed them leaving, and they killed or raped somebody," he recalls. "And then the computer showed they got home when they were supposed to. And so immediately

I started looking for something that had better tracking ability, to be able to track them wherever they went."

In the early 1990s, Nimer received a call from Bob Martinez, who had just served a term as Florida's governor. Martinez was working with some former engineers from Westinghouse, the first company to get a grant from the federal government to design electronic monitoring systems. They were now in the process of launching a company called Pro Tech, the first company to take seriously the idea of building a monitoring system around GPS. Martinez asked Nimer if he might be interested. In 1994, Nimer showed a prototype device at a convention and was nearly laughed out of the room.

"I thought it had a future, but everybody and their brother thought I was crazy, that it was absurd, it would never work, and that the ACLU would put a stop to it," he says. The general problem, as most detractors saw it, was not that GPS monitoring would be soft on crime, but that the idea of monitoring someone around the clock was too impractical. Nimer not only championed the idea, he also wanted to use it on more hardened criminals, whereas most monitoring systems were used for nonviolent offenders on parole or probation. "My conception was that we had to use it on the worst of the worst," he says. "I didn't want to use it on nonviolent drug offenders. Why would I waste my time monitoring their whereabouts? Violence was the key issue."

As GPS tracking of offenders became more common, it was increasingly associated with sex offenders, especially after Florida's passage of "Jessica's Law" in 2005, which mandated lifetime GPS tracking of all convicted sex offenders. Variations of Florida's law have since been passed in more than half of the nation's states. By 2009, more than half of all offenders in the U.S. subject to GPS tracking were sex offenders. And sales of GPS monitors were on the rise, amounting to a third of all monitoring systems in use, with more added each year. The number of people wearing legally mandated GPS trackers in the U.S. is difficult to quantify with any certainty, but probably hovers around 80,000 people each day.

In the context of the overall population of three million people in prison or under some sort of legal supervision, that is a relatively small number. Many European countries typically have fewer than 100 people subject to GPS offender monitoring. The exception is Great Britain, particularly England and Wales, where the Ministry of Justice is instituting plans to track as many as 75,000 people on any given day. The push there is to use GPS tracking as a stand-alone measure, obviating the need for government services such as probation, while offering lucrative private contracts for manufacturing the devices and monitoring prisoners. "I always feared that in England and Wales, the government would become so enthusiastic about electronic monitoring that they'd feel they could do without a probation service," Nellis says. "And that's kind of what's happening."

GPS tracking in Britain received a boost with the launch of a company called Buddi in 2005. Sara Murray, Buddi's founder, had already made a small fortune with an online insurance company she created. Still in her thirties, she had no experience with electronic monitoring, but found inspiration in a few panicked parental minutes. "I literally lost my daughter in a supermarket, that's how we got going," Murray says. "I turned around and she wasn't there. That was just a heart-stopping moment. I found her really quickly, but I thought, this is ridiculous, I need to give her something in case she runs away again."

Buddi first made waves by marketing a tracker for use with dementia patients, and later worked with psychiatric hospitals to fit certain patients with trackers. Buddi worked with police in Hertfordshire on a pilot tracking program, mostly used on nonviolent repeat offenders. It was considered such a success that police around the country began similar programs. By 2014, most of the forty-two police departments in England and Wales were experimenting with tracking, and nearly all used Buddi's tracker, setting the stage for the Ministry of Justice's ambitious plan to expand tracking in both countries.

Buddi is expanding its scope and making inroads into the American tracking market. "All the GPS tags that existed before we came to market, the offenders have to plug themselves into the wall for eight hours, and the attitude in America is, 'Well, so what—you know where they are, right?' " Murray says, laughing. "In Europe the attitude is, 'You must be joking—what about their human rights? You can't make them do that!' "

Buddi is also exploring the possibilities of a cottage industry that has arisen within GPS tracking, with companies marketing intelligent mapping software that notices behavioral patterns and raises red flags based on the movement of people wearing tags. "Most of our patents are around behavior analysis and using algorithms and heat tracking," Murray says. "What is normal behavior, and what looks like abnormal behavior, and is that OK or not? Police would look at data from someone's day and say, 'Oh, that looks odd.' "

Murray sees this kind of intelligent tracking as a weapon against recidivism. "What we have seen are things like an offender who was a prolific burglar who stopped going out at night," Murray says. "He's in at night, so that's a massive step forward, but if you look at the patterns, you see that every Thursday or Friday, he's hanging around a street corner for a slightly unusual period of time. So police go and watch the corner and they catch him dealing drugs. He's changed his m.o. from being a burglar, because he knows he'll get caught. You can pick up on it quite easily in a pass of the data, because if you watch a whole series of data you'll see very quickly that people don't hang around in outside places much, unless there's, say, a bike being stolen."

Buddi is also at work researching a tracker that can detect the presence of drugs or alcohol. That puts Buddi squarely in the tradition of the Schwitzgebels, who also imagined an alcohol-detecting tracker. Like them, Murray understands that location knowledge is just the beginning, a foundation for learning much more about someone's behavior. What is less clear is how much the ethic of a company like Buddi has in common with those groundbreaking brothers. They saw trackers purely as a social good. Is that a fallacy today? Was it one then?

Around the time I was trying to track down Robert Gable, né Schwitzgebel (both brothers changed their surname to Gable), I heard about the development of the world's smallest GPS tracker, tiny enough to attach to a bee. I imagined an entire swarm of tracked bees, and trying to make sense of the patterns and lines their movements would leave on a digital map. If every member of the swarm is tracked, that is a form of egalitarianism, right?

When I reached Robert on the phone, Ralph was in poor health. (He died in 2015.) But Robert was in good spirits, describing the playful aspect he and his brother believed was important to their idea of positive reinforcement through tracking. The idea was to keep things unpredictable, subtly engineering life to positively reinforce behavior. Robert recalled the kid whose tracker located him as showing up on time for his gas station job. The brothers rewarded him with a week's worth of limousine service between home and work. Every day, he would get dressed up "like a Wall Street banker" for the ride, changing into his work clothes when he arrived. "It's theater of the streets, it's fun, and that's the reinforcement," he said.

Robert still thinks GPS tracking could benefit from a street theater element, encounters engineered by people doing the tracking based on knowledge of where the tracked person is at that moment or is likely to be later. Reinforcing positive behavior with rewards would be part of the rehabilitative aspect of tracking, which he thinks has all but disappeared from the tracking model. (In 2007, he compared seeing the tracking industry develop to "watching a child grow up retarded because of being misunderstood.")

Not that he has any illusions. As offender monitoring grew in the 1990s and 2000s, Robert and Ralph would occasionally meet with company representatives to communicate these ideas. "They were not interested," Robert says. "It's a tough, Wild West type of business. They weren't interested in any sort of flaky positive reward system. They're just trying to protect the public. A lot of them are ex-police officers, and they just don't have a social work mentality." Ralph even tried to rally interest in a plan for groups of parolees to track one another, looking out for potential bad behavior and intervening—shades of Ralph's original *West Side Story* epiphany. "That didn't take off, and I don't know why—maybe people just don't want to reward criminals."

Robert knows a system devoid of any punitive qualities would be unworkable and undesirable. But when he talks about the type of system he and his brother first imagined decades ago, he reveals an attitude toward privacy similar to Liang's. "Yeah, it's kind of Big Brother-ish, and it depends on how you feel about privacy," he says thoughtfully. "I would say I don't care about privacy. Privacy is only needed when you have a punitive society. If you have a culture such as what Skinner wanted, you don't need privacy. We hide when we're going to be punished. We don't hide when we're going to be rewarded. So the stress on privacy is really a critique of punishment."

The GBAS interference at Newark Liberty was stronger than usual on August 4, 2012. It was also more constant, less transient—not the elusive blips generated by GPS jammers in vehicles speeding through the Turnpike. Police placed a call to the FCC, and within an hour an agent from the FCC's New York office was on the scene, with signal-tracing equipment. Whatever it was, the signal was broadcasting on parts of the frequency band reserved, all over the world, for radionavigation satellites, including GPS.

The agent followed the noise down a service road that paralleled the Turnpike. As he approached Guard Post India, an entry point to the runway area, the signal grew stronger. It was blasting from a red Ford F-150 pickup truck. Airport police swarmed around it. They found the driver inside.

His name was Gary Bojczak. He worked for a company called Tilcon New Jersey. Bojczak admitted that he had a GPS jammer inside, and that he'd installed it to block his company's fleet tracking system. He surrendered the device to the agent, who shut it off and confirmed the interference had stopped.

The airport authority had hired Tilcon to manage a $26 million project to rehabilitate and upgrade runway 44-22L and the several miles of taxiways surrounding it. The jamming problem had, in effect, jumped the fence. An employee of the company hired to modernize the runway had used GPS to render useless the GPS system that was a key part of that modernization. There was never any danger of Bojczak crashing a plane, but there was a clash of sorts on the day he was caught. The desire to use GPS to pinpoint location exactly—enough to guide a blindfolded jet airplane—collided with one man's desire to thwart that power and vanish completely.

Using a calculation system related to the laws the agency accused Bojczak of breaking, the FCC declared his offenses merited a fine of $42,500. They decided to cut him a break, lopping off about $10,000 because Bojczak had surrendered his jammer voluntarily.

CHAPTER EIGHT

Return from Mid-Ice

The blast of energy was so large, so overwhelming, that if it could have been somehow channeled into turbines and generators, it could power a city the size of Los Angeles—or the entire state of Oregon—for a year. That is one way to comprehend the seismic power of the earthquake that hit Japan in March 2011—an earthquake that redistributed the planet's mass in a way that made it spin faster, so that every day on Earth is now 1.8 microseconds shorter. The Earth didn't just move—it *moved.*

The residents of Oregon may soon experience this kind of earthquake, which scientists call a mega-quake. The Pacific Northwest mega-quake could be a mirror image of Japan's: a massive seismic disturbance in the ocean, not far from the coast, generating both a quake and a tsunami. The last mega-quake in the region occurred in 1700. The estimated recurrence rate is every 300 years. In Southern California, the San Andreas Fault is overdue for a mega-quake that will likely be slightly less intense than what the Pacific Northwest has coming, but still brutally powerful.

The San Andreas is perhaps most famous as the cause of the mega-quake that devastated San Francisco in 1906, but this northern section of the fault tends to shift incrementally, gradually relieving pressure. (While the San Andreas remains a concern, Northern California probably has more to fear today from the Hayward Fault, which lies beneath the major population centers—San Jose, Oakland, and Berkeley—of the East Bay.) The real Andreas-dread is felt in Southern California, where the fault can remain relatively stable for many years, accruing stress, until the land impulsively decides to rearrange itself. The southern tip, stretching from Los Angeles to the Salton Sea, experiences a mega-quake every 150 years, on average. It has lain dormant for nearly 300.

It is impossible to predict with any certainty exactly when an earthquake will occur. "That's been a failure every time it's been tried, to my knowledge," said Larry Young, a scientist who supervises the GPS Systems Group at NASA's Jet Propulsion Laboratory in Pasadena. We were in his office, on the second floor of a barracks-like building at JPL's headquarters, which sits against a hillside in a bedroom community a couple of miles south of the Angeles National Forest. The San Andreas skirts the northern edge of the forest, some 50 miles away.

When combined with traditional seismic monitoring equipment, GPS can give a region a crucial extra few seconds or minutes of warning just as an earthquake starts. It can also, in a way that seismic tools cannot, help scientists understand how an earthquake deforms the land, long after the quake has hit.

JPL is the site of some of the most sophisticated GPS research, much of it applied to tracking spacecraft, including the Mars rovers, which the facility designs and builds. But a good portion of its work for the last forty years has been in harnessing GPS to improve earthquake monitoring. Young was among the first scientists to explore the possibilities of building GPS receivers sensitive enough to detect seismic activity, never thinking that GPS would become a lifelong pursuit. "I thought we'd be through with GPS twenty years ago," he said.

GPS can register movements of the earth that are quick and dramatic, as well as those that are slow and protracted. The story of how GPS came to serve this function begins with the heretical idea that the earth moves at all.

The German scientist Alfred Wegener was not the first person to notice the curious relationship between the east coast of South America and the west coast of Africa. On a map of the world, the two coastlines appear to match, as if someone has ripped a page jaggedly in two. The first to go on record about this mysterious geography was probably Francis Bacon in the early sixteenth century. Others who noted the convergence included the Flemish cartographer Abraham Ortelius, in the same era as Bacon, and the French scientist Georges-Louis Leclerc, in the eighteenth century. Aside from idle speculation that the two continents may have once been joined together, none of these men pursued the theory.

Wegener, a geophysicist, grabbed hold of the idea. A lecturer in meteorology, astronomy, and physics, Wegener had a stern, brooding demeanor and "disliked any kind of stage management, and, as a strong individualist, did not care for large organizations," his friend and colleague Johannes Georgi recalled. "On the other hand, he much preferred well-directed work by one man and a few tried comrades." Wegener loved the pursuit of an elegant idea, the dogged determination to follow a concept to the ends of the earth. Which, in more ways than one, is what he himself did. "It was impossible to escape the attraction of his direct, simple, natural personality," Georgi remembered.

One day in the winter of 1910, Wegener was paging through a friend's atlas when the image of those two ocean-crossed continental lovers caught his eye—especially Brazil, which looked like it could fit snugly into the Gulf of Guinea. He cut the world's continents out of the book, played with them like pieces of a child's puzzle, examined topography, and found other intriguing matched pairs, such as the Appalachian Mountains and the

Scottish Highlands. He eventually found a way to fuse the continents into one giant land mass that he called Pangaea. Several months later, he discovered a scientific paper that detailed identical fossils of plants and animals from both sides of the Atlantic. It seemed logical to conclude that they had evolved on the same piece of land.

Georgi maintained that the seed had been planted in Wegener's psyche a few years earlier, during a two-year scientific expedition to Greenland that Wegener joined in 1906, at the age of twenty-six. "He could see more profoundly than others into the intricate connections of the atmospheric machine, and—though trained as an astronomer—not by mathematical calculation, but by intuition," according to Georgi. "He saw obscure causes and effects with the inner eye, without consciously taking a step forward." Wegener compared his observations with those he gathered on a subsequent Greenland expedition in 1913, as well as others he undertook later in his lifetime. These treks, which traversed hundreds of miles and involved days of trudging through snow and ice, required a physical fortitude that Wegener somehow maintained, despite being a chain-smoker.

Wegener first publicly discussed his theory a year or so before his second Greenland expedition. While recuperating from wounds suffered while fighting in Belgium during World War I, he began drafting a manuscript describing his theory, which he now called continental drift. Wegener posited that 200 million years ago, Pangaea began to separate along rifts, creating valleys that widened into oceans as the gaps grew larger, the continents moving apart as they plowed through the oceanic crust. *The Origin of Continents and Oceans*, published in 1915, went largely unnoticed, an outsider's ravings too crazy even to be panned.

A revised edition published four years later attracted some attention, nearly all of it negative, but it was after the appearance of an English translation, in 1922, that the invective flowed. A paleontologist named E. W. Berry savaged the continental drift theory as "a selective search . . . ending in a state of auto-intoxication in which the subjective idea becomes an objective fact," a critique that accused Wegener of indulging in both onanism and mystification. The American geologist Bailey Willis called continental drift a "fairy tale" that "encumbers the literature and befogs the mind of fellow students." Philip Lake, a British geologist, accused Wegener of "not seeking truth . . . blind to every fact and argument that tells against it." Rollin Chamberlin of the University of Chicago questioned whether one could take Wegener seriously and still call oneself a scientist. "If we are to believe Wegener's hypothesis," he averred, "we must forget everything which has been learned in the last seventy years and start all over again."

The prevailing theory to explain the mirror-image coastlines was that land bridges connecting these continents had at one time collapsed and crumbled into the oceans. Most geologists believed that mountains were created by the contraction of the earth's surface due to cooling, which Wegener derisively compared to the way "a drying apple becomes wrinkled by folds on its surface through the evaporation of its interior." Wegener believed that mountains were the result of the leading edges of continents colliding, forcing the land to pile up into mountain ranges from the constant pressure. India, he argued, shoving itself into Asia at a rate of about 5 centimeters per year, had created the Himalayas, and similar processes had formed the Sierra Nevada and the Andes.

Wegener continued to publish updated versions of *The Origin of Continents and Oceans*, incorporating new material that argued in favor of continental drift. But with a few exceptions, the world's geological community refused to let Wegener into their club. "It is inconceivable that masses of continental size should move over such large arcs and preserve their outlines of either coast or continental margin intact," Berry told an audience, adding, "I have a feeling that it is as futile to discuss the interior of the earth until we have more facts, as it is to attempt a 'scientific' proof of a future life, or the divine inspiration of the Pentateuch."

Over the next few decades, new methods and technologies began to chip away at the mass rejection of Wegener's theory. There was the confirmation, in the early 1950s, of the existence of the Mid-Atlantic Ridge, a mountain range on the floor of the Atlantic Ocean whose contours appeared to match those of the continents on either side of it. New techniques for dating land suggested that the ocean floor was younger than the continents and that newly created land moved outward, a process called seafloor spreading. The end of the 1950s saw the development of geochronology technologies, which can determine the ages of magnetized strips on the ocean floor, adding to our understanding of its age.

The strongest vindication of Wegener's theory, and a useful critique of it, appeared in 1960. An American geologist, Harry Hess, proposed that the mid-ocean ridges are where the seafloor is born. Magma bubbles up from rifts, building up on either side of them to form underground mountains. As the piles grow, the older matter slides down and away from the slopes, at a rate of about one centimeter per year. The youngest matter is always closest to the rifts. The constant influx of magma exerts constant pressure on the layer of the earth's interior called the mantle, which begins a few miles under the ocean floor. The pressure keeps the land in constant motion, causing the seafloor to spread. The continents get carried along for the ride.

This is what Wegener had gotten wrong. The continents themselves do not drift. They are perched on the Earth's lithosphere, the planet's outer shell, which is divided into hard, rigid plates. Driven by molten rock and the heat radiating from the residue of radioactive materials, it is the plates that are in constant motion.

The continental drift hypothesis became the theory of plate tectonics. It posits the existence of two types of plates: continental plates and denser oceanic plates. When two continental plates collide, they deform the land and create mountains. The Himalayas were not caused by India exerting pressure on Asia, as Wegener thought, but by the collision of the Indian Plate and the Eurasian Plate. Continental movement was a byproduct of plate movement.

When a continental plate and an oceanic plate converge, the heavier oceanic plate is pressed back down into the earth, forming an area called a subduction zone. This collision may produce mountains; it may also create volcanoes and trigger earthquakes. The San Andreas Fault—the line where the Pacific Plate (oceanic) meets the North American Plate (continental)—is responsible for California's Santa Cruz, San Gabriel, and San Bernardino Mountains, and is notorious as a bringer of earthquakes.

By the late 1960s, the plate tectonics theory was accepted by most of the scientific community. To further understand the processes involved, geophysicists needed tools Wegener lacked. This was a low-level quest, overshadowed by the giddy reach for the stars that characterized the Apollo missions. While so much of the world turned its gaze upward, dreaming of extending humanity's reach into outer space, these scientists trained their sights on the ground underneath their feet, hoping to reveal the planet's innermost secrets.

In August 1969, less than a month after Apollo 11's historic landing on the moon, sixty-five geophysicists gathered at Williams College, in Massachusetts, for a meeting convened by NASA to discuss the need for improved techniques to measure positioning, velocity, and acceleration of points on the earth's surface. The report produced by this meeting—*The Terrestrial Environment: Solid-Earth and Ocean Physics*, known today as the Williamstown Report—addressed the problem in the one-world rhetoric used by the nascent modern environmental movement (the first international Earth Day was less than a year away): "The planet earth is the <u>only</u> home for the human race for at least several centuries to come. . . . From a social point of view, it hardly needs emphasizing that the mechanics of the earth are integral to the problems of man's environment, which are receiving such prominence in today's world."

The report placed a premium on advancing the study of plate tectonics, especially by using a technology called very-long-baseline interferometry. VLBI was designed primarily to maximize the power of radio telescopes, giant antennas that pick up radio signals emitted by celestial objects such as quasars and black holes. VLBI calculations involve pointing two or more telescopes, located on opposite sides of the world and synchronized with one another by atomic clocks, at the same celestial body, and noting the different times the same signal reaches each telescope. The combined data yields a radio image that is richer and more detailed than that provided by a solitary radio telescope.

One useful byproduct of using VLBI for astronomical observations is that one can also determine the exact distance and vector between any of the linked radio telescopes, even if they are thousands of miles apart. The measurements are so precise, in fact, that scientists can perceive very small changes in distance. When a plate moves, so does everything above it on the surface, including radio telescopes. VLBI, along with another technology called satellite laser ranging, which measures distance by bouncing tiny light pulses off a satellite, offered the first real chance to observe the movement of tectonic plates.

Much of the early VLBI research was done at MIT, and partially funded by the U.S. Air Force Geophysics Laboratory, which considered VLBI a potentially valuable Cold War military asset for precise positioning. Should the U.S. ever decide to launch a missile at a Soviet nuclear silo, the Air Force wanted the missile to hit within a few meters of the top of the silo. The earliest proponents of VLBI were probably two MIT physicists, Irwin Shapiro and Alan Whitney. Another brilliant and occasionally pugnacious MIT physicist, Charles "Chuck" Counselman, contributed groundbreaking work. On the day Apollo 15 landed on the moon in 1971, Counselman and another MIT professor tracked the position of the astronauts' rover, relative to the module, using VLBI measurements of the radio transmissions. After following the rover as it made a four-kilometer jaunt around craters named by the astronauts Spook, Buster, Halfway, and Flag, the final position calculation was off by only 30 meters.

Young and his colleagues, like Counselman and his East Coast cohort, experimented with semi-mobile VLBI solutions, taking measurements by driving trailers loaded with equipment, with large dish-shaped antennas on top. Two teams at two different points, often separated by several miles, would take simultaneous VLBI readings and compute the distance between the two. By doing the same thing at a future date, they could note how plate movement had changed the distance between the points.

An early earthquake prediction using VLBI was underscored by the scientific community's reaction to the apparent swelling of the earth along the San Andreas Fault, around 60 miles north of Los Angeles, an 18-inch upward surge nicknamed the Palmdale Bulge. In 1978, hundreds of scientists descended on the area, with many supporting the hypothesis that the bulge was caused by the fault crushing rock and making water in the ground rise up, a harbinger of a giant quake. Among the supporting evidence for this theory were VLBI measurements. "The Palmdale Bulge just seemed ripe for the picking," Young said, recalling that it even made the cover of *Life*, pictured with a JPL scientist—"much to his later chagrin." Over the next few years, evidence for the bulge theory began to decline, while the topic engendered fierce—and sometimes nasty—debate among scientists. Eventually, the last bulge holdouts had to admit defeat. It wasn't even clear, in retrospect, whether the swelling was visible. "If we'd had GPS in those days," one Caltech seismologist later pointed out, "we could have said yes or no."

The Palmdale Bulge fiasco exposed an undeniable truth about VLBI: it was an unwieldy technology, and a lot could go wrong. So when GPS satellites did begin to multiply in the post-bulge era, the plate tectonics crowd saw a useful alternative. "We realized, hey, the Air Force is going to put up a satellite broadcasting an enormous amount of energy, spread over half the earth," Young recalled. "We can use those instead of quasars!"

But this epiphany came with its own challenges. The advantage of using a quasar as a beacon is that it is so distant from Earth that it appears to us as a stationary object—"a fixed lighthouse," as Young put it. No matter how fast the quasar might be moving in relation to us, the change is trivial compared to the vast distance between us and it. GPS satellites were easier to locate and listen to and high enough to provide decent coverage, making it possible to measure the distance between widely separated points, but they were in constant motion, close enough to us to be considered moving objects. The slightest deviation from their predicted motion, or the smallest amount of delay in receiving the signal—from ionospheric interference or other complications—would affect readings.

Though their exact methods differed slightly, both coastal teams reached the same conclusion and devised the same solution. They would build dual-frequency receivers, which would begin to enter the mainstream in the 1980s, via the work of Chuck Counselman and his colleague Sergei Gourevitch, Charlie Trimble, Javad Ashjaee, and others. The civilian GPS signal was freely available, but nowhere near precise enough to detect changes in plate movements. The military signal's code was impenetrable, but the signal itself could be picked up. Also, the ephemerides of the satellites—the details of their orbits—were publicly available. By knowing the positions of the satellites at any given moment, a measurement of the military signal's phase could be used as ranging information.

What the scientists discovered—and what Charlie Trimble

and others would build upon—was that these phase measurements could potentially yield location data that was *more* accurate than what even the military code could yield. Any radio signal is a series of waves, and every signal has a wavelength, the distance between the peaks of those waves. The wavelength of the GPS carrier signal is much smaller than the wavelength of the information signal. Like the difference between centimeters and millimeters on a ruler, this leads to very precise measurement. The more the scientists examined GPS, the more they marveled at its untapped potential. "It just seemed like the Air Force was using GPS in the crudest possible way," Young said.

The implications for plate tectonics research were profound. Counselman, at MIT, made the first big public splash. In 1979, he delivered an after-dinner speech at an annual convention of geodesists and geophysicists in Canberra, Australia. For many in the audience, his speech contained the ravings of a lunatic. He was there that night to tell them that he, Chuck Counselman, could measure the distance between two points separated by hundreds of miles to within a centimeter—and he could do this by harnessing the power of the Global Positioning System.

The response was Wegenerian. Counselman had expected astonishment, laced with skepticism. What he got was open hostility. One NASA scientist called him a snake oil salesman. To Counselman, it felt as though he'd stood up and casually mentioned he'd found a cure for cancer while tinkering in his basement.

Counselman made it clear that he planned to commercialize his GPS hack by designing a receiver for the surveying market. That put him in an awkward position with the military, which was partially funding his research and which now discovered that Counselman was essentially telling the world that everyone could have military-grade GPS precision.

In February 1981, Counselman was summoned to the Pentagon. He explained that his device could measure the distance between two points with a margin of error of a few millimeters, using measurements gathered over a period of one to two hours. He was told the best the Air Force could do, using data gathered and averaged over a few weeks, was about 30 centimeters. "My claimed accuracy was ten thousand times better than what the Air Force was claiming at the time," Counselman recalls. "My explanation was met with utter disbelief, especially by the chairman of the meeting, who was a high-ranking Navy official."

"What saved our bacon at the time was that the techniques that we used could not be done in real time," says Thomas Herring, an MIT geophysicist and one of Counselman's former graduate students.

By 1983, Counselman and Gourevitch were putting the receiver, which Counselman called the Macrometer, through a series of public tests for government agencies involved with surveying and mapping (as well as for clandestine representatives from two intelligence agencies). Gathering data with the Macrometer required programming a special desktop computer with the orbital data for the satellites (the constellation then numbered five). The raw data from a day's observations was transferred onto a cassette tape for processing at another office. A report on the findings declared that the distance computations taken by a Macrometer "were consistent at a level that was significantly smaller than expected on the basis of prior estimates of the uncertainties of the terrestrial measurements."

But how would the Macrometer fare against its nemesis, JPL's dual-frequency precision GPS receiver, rakishly named the Rogue? "He was a competitor," JPL's Larry Young said of Counselman, leaning back in his chair and smiling, "with a very inferior device that did not match the stellar performance of JPL's."

"This statement, although I am sure Larry would make it, is not strictly true," Herring says. "The difference was that the JPL Rogue used the military codes that were available at the time, although the military had always indicated that they would not be available to the public when GPS went operational. Counselman's receiver did not need those codes—that was its innovation—but the data accuracy was compromised by a lower signal-to-noise ratio."

For three weeks in January 1984, the two receivers—along with a device built by Texas Instruments, the TI-4100, the one other geodetic-quality receiver on the market—engaged in a "shoot-out" sponsored by the federal government. The three teams were instructed to measure the correct vector between far-flung points, such as Fort Davis, Texas, and Owens Valley, California; whichever came closest to the correct value, based on previous surveying, would be the winner. Tempers sometimes flared in the field. "One of the highlights of the experiments," Young recalled, "was some contention between [the Texas Instruments group] and the MIT group, and you know the personality there," Young said, apparently referring to Counselman. "The MIT group said the TI-4100 was emitting radiation that jammed their receivers. One of the principals"—Counselman again—"ended up calling one of the other principals a fat, waddling pig."

Beginning in the late 1980s, the number of GPS receivers measuring plate movement began to multiply around the world. The next step was to design systems that could harness these ultra-accurate receivers specifically to study earthquakes. In an impressive display of comity, East Coast and West Coast came together to explore the new frontier of precision GPS. Yehuda Bock, a member of Counselman's group, decamped from MIT to the University of California San Diego. In the spring of 1990, Bock and a contingent from JPL—including Larry Young's team—implemented the California Permanent GPS Geodetic Array. Five receivers, scattered around sites in Southern California, took a measurement every thirty seconds. If seismic activity occurred in the region, the array would record it.

The network proved its worth less than two years later, when it captured a 7.3-magnitude earthquake that struck a remote section of California's Mojave Desert, near the small town of Landers. Although it was the strongest earthquake to hit California in years, the location limited the damage and injuries. Less than two years later, the network captured the 6.7-magnitude quake that hit Los Angeles, along a previously unknown fault line in the Northridge section of the San Fernando Valley, which killed fifty-seven people and injured more than 5,000. (Among them was Bock's future wife, who lived very close to the epicenter.) The property damage—estimated at between $13 billion and $40 billion—made it one of the costliest natural disasters in the nation's history.

The Northridge earthquake provided the political incentive to expand the network, which soon grew to include 250 continuously operating GPS receivers, funded by the federal government and private foundations. The proliferation and success of similar networks inspired UNAVCO, a nonprofit university-funded consortium headquartered in Boulder, Colorado, to build 1,300 stations throughout the western continental U.S. and Alaska, collectively known as the Plate Boundary Observatory, whose stated aim is to provide real-time monitoring of the "rapidly deforming" boundary between the North American and Pacific plates.

As these networks have multiplied, a parallel movement has grown to complement their efforts. The type of precision plate tectonics research demanded was beyond what the basic GPS infrastructure—the monitoring stations that track the satellites, the crews in Colorado uploading new instructions to the GPS satellites—could provide. To obtain perfect knowledge of the world, to observe and record the smallest movements of its massive plates, the scientists essentially needed to have a perfect understanding, at every moment, of the satellites' orbits.

In the early 1990s, when the initial five receivers in the Permanent Array were going up, scientists began to build a global civilian GPS network, not linked to the U.S. military, specifically to improve knowledge of the satellites' orbits. Today, the International GNSS Service, a mostly volunteer consortium of scientists, oversees hundreds of geodetic-quality tracking stations around the world. Several data and analysis centers compile and process the data. The output is information about the satellites' orbits that scientists can use to strengthen GPS readings.

The interaction between the networks that monitor plate movement and seismic activity, and the IGS network, neatly illustrates the feedback-loop ubiquity of GPS today. A network of high-quality GPS receivers keeps watch on the satellites, so as to improve the performance of the geodetic networks of GPS receivers. GPS makes GPS possible.

These networks came together to provide a complete picture of the world Wegener described. But once they were in place, scientists realized that this was only the beginning of their usefulness. They are the reason GPS can help us get a jump on the next mega-quake.

The observation and measurement of earthquakes did not, of course, begin with GPS. Large networks of seismic stations have long existed in California and elsewhere. But the effect of GPS on geophysical research has been so transformative that, like Wegener's theory a century ago, it demands an intellectual reorientation. The world looks different.

In November 2002, Alaska experienced an earthquake with a magnitude of 7.9, the largest tremor to hit the interior of the United States since the advent of modern earthquake recording. (The epicenter was in a remote section of Denali National Park, so injuries were minimal.) Geodetic-quality GPS receivers were in the area, but although they could detect small movements in tectonic plates, they were of limited use when gathering earthquake data. They took a reading only every thirty seconds, what geophysicists call "survey mode." A small earthquake can last as little as ten seconds, and a large quake rarely exceeds one minute.

A typical reading of an earthquake from a GPS receiver might be one sample that catches the middle—maybe two if the quake is big and the timing right.

But land surveyors also use precision GPS receivers, and some of them happened to be taking readings in southern Alaska during the Denali earthquake. Although they were over 800 kilometers away from where the rupture started, the seismic waves were noticeable in the GPS data. Their equipment took a reading once every second. Kristine Larson, a geophysicist at the University of Colorado Boulder, was curious what their GPS data would show. "I didn't know how big the waves would be, and I couldn't get a seismologist to tell me," she says. "Because every time a big earthquake had happened before GPS came around, the seismometers would saturate near the earthquake." This has always been a problem with most seismic instruments. Large quakes overwhelm them, rendering them temporarily useless. In these situations, the earthquake's true size is inferred, using other methods, after the quake has hit.

When a large quake hit Hokkaido, Japan, the following year, Larson was able to use geodetic receivers that were much closer to the rupture, providing the kind of moment-by-moment traces of a large quake that nobody had ever seen. "I showed these to a professor I had taken classes from in grad school, and I said, 'Look, I have these measurements near the big earthquake in Hokkaido—this is what they look like,' " she recalls. "And he told me I was wrong! He said, 'The earth doesn't do that, Kristine.' I still remember that—'The earth doesn't do that.' I was kind of shaken up by it. I talked to a colleague. I said, 'How could he tell me that this is wrong? I didn't make this up. I'm measuring it.' And he said, 'Kristine, he has never seen an instrument on scale during a large earthquake. It always saturates. He doesn't know what they look like.' "

The data looked odd to the professor, partly because he was accustomed to data derived from seismic instruments several thousand kilometers away from the epicenter—but mostly because he had never seen this kind of data at all. Seismic instruments measure the velocity and acceleration of a wave. What they cannot measure—but which GPS can—is the land displacement caused by an earthquake, how the position of a particular point after the quake is different from its original position. The velocity/acceleration data is very noisy—it reflects the earth bouncing up and down very quickly. Land displacement data is very smooth: the land starts in one place and shifts to another. The professor—and every scientist who had only studied earthquake data from seismic instruments—had never had access to this data in any form.

Today, GPS data from the Plate Boundary Observatory and other networks transmits continuously and freely on various wireless networks and the Internet. Before advances in disk space, processing power, and telemetry, gathering and analyzing data was a labor-intensive process. The Denali quake happened at a time when the problem was not receiver technology—all geodetic-quality receivers could sample at a rate of one per second, or even greater—but rather what to do with all the data they collected.

"In the mid-eighties, when I was at MIT, we started doing GPS measurements in survey mode," Yehuda Bock said. We were in his office at the University of California at San Diego's Scripps Institution of Oceanography. The building is about a mile from the main campus, perched on a cliff above the Pacific. I couldn't help wondering what would happen to the structure in a megaquake. "You basically went to the field, and you would survey markers over a day, or several days. You come back maybe a year later, and you keep repeating the measurements. Anywhere there was earthquake activity, an active geologic fault, or a plate boundary, somebody was working on doing these surveys on a periodic basis."

Survey mode is good at revealing the slow, steady movement of the earth's surface due to plate tectonics. Continuously operating stations give the data more granularity, showing the steady tectonic motion—the maximum rate is about 120 millimeters per year, and about 50 millimeters per year in California—but also variation in the steady state of the movement. This is especially true when measuring the effect of seismic motion. "During earthquakes, there are large displacements of the ground, up to several meters for the largest earthquakes," Bock said. "And there's also what's called post-seismic deformation, where the earth responds and then goes back to its steady-state operation. This could be happening over several years. Knowing the offsets that happen during an earthquake, and knowing how the earth responds, is fundamental for improving our understanding of earthquake physics and how faults work." Analyzing the GPS data is like putting a time-lapse camera on a skin wound, showing how it heals itself but leaves a scar.

Bock pulled up on his computer a graph that displayed data gathered by a continuously operating GPS station during a medium-sized quake (magnitude 6) that occurred along the San Andreas Fault in 2004, and also showed the data for the nine months following the quake. The earthquake itself registered as waves that dissipated over one to two minutes for this event—the same data a seismometer would record—but there was also something else. "The strong shaking starts and so you have the offset, we call it co-seismic offset, which is permanent displacement," Bock said. "That's where the earth ends up, and you can see that this particular site jumped two centimeters." The data showed that the ground had shifted four centimeters during the quake, and then six centimeters during the several months that followed. "Seismic instruments can measure this," Bock said, pointing to the lines representing the quake itself. "In this period"—he pointed to the months following it—"they measure nothing. With GPS—and *only* with GPS—you can see the shaking as well as the long-term motion."

The ability of GPS to measure displacement in unprecedented

ways is not just a way to apprehend the complex processes earthquakes enact on the earth. They also allow researchers to model the effect of earthquakes on buildings, by attaching GPS receivers to structures built on shake tables—vibrating platforms used to imitate seismic effects. But there are also things that seismic equipment can do that GPS cannot, such as measuring microearthquakes that typically occur after larger seismic events. Bock and his team at Scripps have pioneered the use of "seismogeodesy," combining GPS with traditional seismic equipment. In 2004, researchers used a seismogeodetic approach to study the behavior of the Verrazano-Narrows suspension bridge, the starting point for the New York City marathon, during a forty-hour period that began ten hours before the race's start. Two stations, each containing a GPS receiver and an accelerometer (a major component of seismic stations), were established on the upper deck of the bridge, one at mid-span and the other 150 meters away.

The setup allowed for an astoundingly subtle assessment of the bridge's vertical and horizontal movements. The first station encountered by the runners leaving the Staten Island starting point showed the bridge's deck beginning to sag under their weight, twelve minutes into the race, reaching its "maximum deflection" of 350 millimeters between five and six minutes later. The bridge required about the same amount of time to bounce back to about 70 percent of its original position, and twenty-seven minutes to make a complete recovery. The second station showed a slightly lower maximum deflection, and a recovery that lagged about two minutes behind that of the first station. The stations recorded smaller horizontal movements of the bridge, between 10 and 30 millimeters, probably due to a combination of a cross breeze and an uneven distribution of runners in the lanes. At the same time, as the bridge sagged and recovered due to the weight of the runners, the vibrations caused by thousands of shoes impacting the pavement resulted in a vertical displacement of less than one millimeter. The motion was different from what the bridge normally experiences, but the results showed that it was well within the safety limits of the bridge's design.

This kind of data helps scientists understand how structures behave in earthquakes—and also illustrates the value of seismic instruments and GPS working in tandem to measure earthquakes. The vibrations excited by the runners had large accelerations but led to a small overall displacement—data best captured by the accelerometers. The gradual deflection of the bridge as it responded to the runners' load—slow movement, significant displacement—was ideally suited for GPS.

On the other side of the Pacific, too far to see from Bock's office window, the two plates that created the San Andreas were enacting another movement in their eons-old pas de deux. Like a right hand wrapped around a left-hand fist, the North American Plate surrounds much of the Pacific Plate, skirting the Aleutian Islands and Russia, the hand's index finger meeting the Philippine Sea Plate off Japan's southern coast. Several miles off the northeast coast of Japan, along the Pacific Ocean's ring of fire, the Pacific Plate is moving west, thrusting itself under the North American Plate to form an area called the Japan Trench, at a rate of about one millimeter every four days. The north of Japan lies on top of the North American Plate.*

On the afternoon of March 11, 2011, a compression of the Pacific Plate caused the North American Plate to rise with tremendous force, triggering an earthquake 17 miles underground, with an epicenter about 43 miles offshore. Thirty seconds after the first tremors, the Japan Meteorological Society gave the quake a magnitude rating of 7.2, and before that first minute was up, the agency had issued an emergency response alert that

* Some scientists consider the part of the North American Plate near Japan to be an independent plate called the Okhotsk Plate.

strong ground-shaking was about to begin. Before two minutes had passed, the quake was upgraded to 8.0, which made it the strongest earthquake the region had experienced since the development of seismographic tools to measure quakes.

Twenty minutes later, a few minutes after the initial tremors subsided, the magnitude was finally upgraded to a 9.0. The scale that measures earthquake magnitude is logarithmic, meaning every step up the scale corresponds to an order-of-magnitude change. This earthquake was 900 times more powerful than originally thought, making it one of the strongest earthquakes ever recorded, anywhere.

The quake itself was just the beginning of the calamity. The energy released by the powerful upward motion of the North American Plate caused a disturbance in the ocean. The enormous amounts of displaced water formed tsunami waves that traveled toward the coast at speeds of up to 500 miles per hour. Japanese authorities knew this was coming, but because of the initial flawed measurements of the earthquake's magnitude, they considerably underestimated the size of the tsunami. A half hour after the first tremors, the first waves, some as high as 20 feet, hit the Sanriku coast, overflowing the seawalls. The tsunami did more damage than the quake. When it was all over, the number of casualties topped 15,000, billions of dollars of property was destroyed, and the area's infrastructure was in shambles, including the Fukushima Daiichi nuclear power plant, where workers struggled to contain a meltdown in three nuclear reactors.

Japan has the most sophisticated earthquake and tsunami early-warning system in the world. Japan also has the world's largest real-time continuously operating GPS network. During this earthquake, both systems performed to code, exactly as they were designed to. But the two systems were not linked in any way, which some scientists now see as a tragic oversight. "They had an excellent system in place—they just didn't use it," says Yoaz Bar-Sever, a GPS expert at the Jet Propulsion Laboratory.

Bock and some of his colleagues at Scripps have pioneered a way to combine the two types of systems into a super-accurate and ultra-fast earthquake monitoring technology called seismogeodesy. When an earthquake starts, the first part of the seismic wave, too small for us to feel, is called the P-wave. Following the P-wave is the much more powerful S-wave, the big jolt. The P-wave travels faster than the S-wave, so as the distance from the quake's epicenter increases, so does the distance between the two waves.

A modern seismic monitoring station usually includes a seismometer, which precisely measures the velocity of ground motion, and an accelerometer, which measures the varying velocity of strong motions. Together, these two tools can usually record the full range of shaking, but not the permanent displacements. However, faced with a mega-quake, they saturate and fail to perform adequately. A magnitude-8 earthquake looks the same to them as a magnitude-9. The seismometer is overwhelmed by the S-wave (it "clips"), and fails to measure it accurately. The response to the Japan quake illustrated this trade-off: a nearly instantaneous detection and alert, but a twenty-minute latency in accurately assessing the size of the quake, and the size of the tsunami a quake that large would generate.

A GPS receiver will not outpace seismic instruments. It will not detect the P-wave, and GPS is a poor way to measure acceleration. (For this reason, smartphones and car GPS units contain accelerometers which help place the moving blue dot on the map.) But a GPS receiver will continue to perform in a mega-quake, and will provide a more nuanced picture of the land displacement in three dimensions.

A seismogeodetic approach involves merging the two technologies. An accelerometer begins the process by detecting the quake, and a precision GPS receiver, immune to the clipping that afflicts a seismometer, performs accurate measurements of land displacement. "It takes the strengths of both systems, and eliminates the weaknesses of both," Bock explains. In practical terms, this means attaching a seismometer to a continuously operating geodetic-quality GPS receiver.

Since so many of these receivers are already in place to measure plate movement, the infrastructure for such a system in earthquake-prone areas already exists. The western United States has about 600 continuously operating GPS receivers now used for earthquake monitoring—operated through a collaboration between the U.S. Geological Survey, UNAVCO, UC San Diego, JPL, and Central Washington University—along with a large network of traditional seismic stations. Bock thinks the networks will some day formally merge under the aegis of the Geological Survey, which has a federal mandate to monitor earthquakes.

For now, Bock's group is the only one practicing seismogeodesy. In the lab at Scripps, technicians assemble solid-state components designed to plug into the receivers at existing GPS monitoring stations. These components contain both an accelerometer and a module to control the GPS receiver and accelerometer. To quicken response time, each module is a self-sustaining node of the larger network. If the station detects movement, the module analyzes the data, based on an algorithm, and transmits the results back to a central location, which can then send out automated warnings that a quake is imminent. Each module costs about $3,000 to build, though Bock expects the cost will drop if they are mass-produced.

By 2015, Bock's group had upgraded fifteen stations along the southern tip of the San Andreas, and worked with UNAVCO on ten stations in the Bay Area that span the Hayward Fault. (The two clusters of stations have already recorded several magnitude-4 earthquakes.) The next step is to implement a similar cluster along the Pacific Northwest coastline, parallel to the offshore Cascadia Subduction Zone, a mirror image of the Japan Trench capable of causing a mega-quake and tsunami event similar to that which hit Japan in 2011.

Using the combined real-time data that Japan's network of GPS receivers recorded during the quake, Bock's team modeled what would have happened if a seismogeodetic network had been deployed in 2011. "We showed that within 157 seconds we would've already known that we had a magnitude-9 earthquake," Bock said. "That's about how long it took for the earthquake to happen. So as soon as the earthquake stopped, we knew it was a 9."

How would this have affected the official response? "The P-wave propagates at, say, four kilometers per second but we get the GPS data in real time," Bock said. "So if you were in Tokyo, where it took maybe two minutes to feel the earthquake, two minutes could be useful, giving time for more urgent warnings to go out." This is about the same advance warning Seattle would receive if a seismogeodetic system picked up a Cascadia earthquake centered off the Oregon coast, 273 miles south of Portland, according to a scenario modeled by a University of Washington geophysicist. Assuming Bock's protocol was in place, two minutes would be enough time to halt trains, divert planes, and give hospitals an opportunity to pause operating-room activities. If the alert automatically went out to mobile phones, social media sites, and navigation programs like Waze, individuals could park cars, take cover, turn off gas, or at least have one lucid moment of abject fear before the deluge hit.

A more detailed understanding of the tsunami moving toward Japan would have given authorities a better indication about how many people were threatened, perhaps encouraging greater numbers to get to higher ground. Officials at the Fukushima nuclear plant would have known that the seawalls near the plant were insufficient for the incoming wave, and possibly taken protective measures. "I'd expect that if they knew twenty minutes earlier, they probably could have done something to mitigate the damage, but I'm not 100 percent sure," Bock said. A recent study by Japanese investigators has indeed confirmed that the underestimated magnitude resulted in a narrower evacuation zone for the impending tsunami.

Southern California will probably not be at risk for a tsunami when the San Andreas mega-quake hits, but the post-quake scenario for a San Andreas Fault rupture is nightmare enough. When I spoke with U.S. Geological Survey geophysicist Ken Hudnut, at the converted house across a tree-lined street from Caltech that serves as USGS's Pasadena field office, he was candid about "the worst of all possible combinations." That would be the quake rupturing gas lines, causing fires to erupt throughout the region, on one of the few days per year when the hot blasts of wind known as the Santa Anas blow down from the mountains—during a period when rain had fallen recently enough to provide vegetation for tinder on the still dry drought-stricken hillsides. "Having a large fire moving across an urban area is a total nightmare, because evacuation routes might not be open after a big earthquake," Hudnut said.

The total fallout of a mega-quake is almost as difficult to imagine as the precise time it will arrive. But fires caused by ruptured gas lines are considered a near certainty. Bock and others have been in talks with local utilities regarding a system that would shut down gas lines automatically in response to the automated warning from the seismogeodetic system. The Geological Survey is working with seismologists to build an earthquake early-warning system for the West Coast, with plans to integrate real-time GPS data to avoid the oversaturation problems encountered during the 2011 Japan earthquake.

"We did this big earthquake scenario for Southern California, back in 2008, called ShakeOut," Hudnut said. "It was based on a southern San Andreas earthquake, and we did all this elaborate simulation, what the engineering effects would be, and studies of the economic impact and social impact, and put it together in a complete package. But we decided to back off from including Santa Ana wind conditions in the scenario, because we knew that would just put it over the top, and people wouldn't want to deal with it."

The GPS signal is so reliable, and dual-frequency geodetic-quality GPS receivers so dependable, that scientists can even extract useful information from the noise in the system. In no field is this more evident than meteorology.

Two types of atmospheric interference bedevil the GPS signal. The problems of the ionosphere, free electrons between 50 and 1,000 kilometers above the planet that delay the GPS signal, are stable and therefore correctable. Those caused by the troposphere, the part of the atmosphere closest to Earth and comprised mostly of water vapor and dry gases, are more difficult to mitigate. By correcting for the ionosphere, meteorologists are left only with the problems of the troposphere. Observing how the GPS signal is affected can tell them much about the concentration of water vapor in the troposphere, data that is very helpful for predicting weather.

For meteorologists, water vapor is the coin of the realm. "Water is a unique substance, in that when it goes from its gaseous phase to its liquid phase it releases enormous amounts of energy, and when it goes from its liquid phase into its solid phase it releases lesser amounts of energy," says Kirk Holub, a scientist who works with the National Oceanic and Atmospheric Administration's Earth System Research Laboratory in Boulder, Colorado. "One way to look at weather is that it is the change of water vapor in the atmosphere. If the water vapor isn't changing much, there's no weather happening."

Water vapor, Holub explains, is nature's way of redistributing energy around the planet. In an unending process, water molecules on the surface of the ocean are hit by solar energy. They evaporate, and in becoming vapor they store that energy. When the atmosphere cools enough for some of that vapor to condense into liquid, much of the energy is released, causing heat. The heat changes the pressure, the pressure results in wind, and the condensed water vapor falls to the earth as precipitation. "The gist of it is that if you want to get the weather forecast right, you've got to get the water vapor right," Holub said.

For most weather forecasting, GPS-derived water vapor data is not very useful by itself. Meteorologists use it to strengthen prediction models that include data gathered through the use of weather balloons and other methods. But there is a way to make more direct use of the data, by computing the amount of water contained in an imaginary column that stretches from the receiver to the limits of the troposphere. Scientists know that within the confines of that column, there is a limited amount of water vapor that can exist in the spaces between air molecules. "There's a theoretical limit of a little under 10 centimeters of water vapor at the equator, where the atmosphere is the thickest," Holub says. "That's all the water vapor you could possibly stuff into the atmosphere. We've learned from watching over the years that for any given location on the planet there's an amount [of vapor] above which there is a very good chance it's going to begin falling back as precipitation."

The most advanced form of GPS-enabled meteorology (Holub calls it the "Holy Grail of weather forecasting") involves a space-based technique called radio occultation. It measures the time required for the signal to travel from the GPS satellite to a GPS receiver on a satellite in low-Earth orbit. That calculation creates a profile of water vapor distributed over several hundred miles.

Even the "wasted" part of a GPS signal can be useful. When the GPS signal reaches a GPS receiver on the ground, some of it travels directly to the antenna, and some of it hits the ground near the receiver and bounces back. The receiver picks up this bounced signal a moment after the direct signal. Kristine Larson, the University of Colorado geophysicist, has developed a method of using these bounces to draw conclusions about the moisture content of the ground, based on the time elapsed between the arrival of the direct signal and the bounced signal. Vegetation with a greater water content suppresses the reflection, making the lag a little smaller in amplitude.

Larson uses these analyses to study California's ongoing catastrophic drought. The work is done using receivers in UNAVCO's Plate Boundary Observatory network. The data is gathered remotely, retrieved from the central archive, processed by the University of Colorado, and released onto the Web.

Larson has also developed a plume sensor for studying volcanic ash. Precision GPS receivers are already useful tools for volcanologists, because they can detect and measure how the ground near a volcano inflates and moves as the underground magma expands. Larson puts the receivers to a different use. When a volcano explodes, volcanic ash degrades the fragile GPS signal, so less of it gets through. Larson measures the strength of the signal when it reaches the antenna, to determine how much of the plume is ash and how much is water vapor released by the volcano.

Larson points out that these measurements have very practical applications. When a volcano anywhere on the Pacific Ocean's ring of fire spews ash, all air traffic in the vicinity is grounded because the ash particles in the air pose a real threat. In 1989, a KLM flight from Amsterdam was preparing to land at Anchorage International Airport in Alaska, a day after Mount Redoubt, around 110 miles south of Anchorage, erupted and spewed ash into the air. Ash clouds do not appear on radar, so by the time the flight crew noticed a curious dark cloud ahead, it was too late to divert. While the pilot tried to ascend out of the cloud, the plane's four engines inhaled the ash and melted it, creating molten silica that covered the engines with a layer of glass, disrupting their sensors. All four failed, and the plane descended more than 14,000 feet before the flight crew managed to restart two of the engines and guide the plane to the airport.

Volcanic ash can affect air traffic over a very large area, with cascading effects. The Mount Redoubt explosion shut down Anchorage and nearby airports for several days, and the drifting cloud complicated air traffic as far away as Texas. Every day, more than 200 large aircraft—many following the great circle routes over the Arctic that link Europe, Asia, and North America—fly over the Aleutian Arc. Anchorage is also a major hub for commercial freight moving between the U.S. and Asia. In 2010, ash from the Eyjafjallajökull volcano in Iceland grounded hundreds of planes over European airspace, stranding millions of travelers for a week or more and costing airports and the airline industry around $2 billion.

Measuring the water in an ash cloud was not exactly part of the original GPS mission statement. "I can assure you," Larson says, laughing, "Brad Parkinson—you know, how could he have thought about something like this? It just wouldn't have made any sense." By the same token, she said, "Who would have thought GPS would be used to tell you where an ice cream store is?"

Each of the thousands of geodetic-quality GPS receivers that dot the globe is like a bright light casting a glow around its little corner of the world. Though gaps remain that keep certain parts of the world darker, there are enough receivers to connect the world into a brightly lit totality. Like no technology before it, GPS provides a way to grasp the enormously complex dynamic processes that constantly shape and reshape our world.

But unless these thousands of receivers march in lockstep, their collective data does us no good. They must do their work based on a shared vision of where each is in relation to others. The model they share is called the International Terrestrial Reference Frame (ITRF), which is maintained and monitored by a federation of 200 government agencies, universities, and research institutions from 100 countries. The ITRF defines horizontal relations—latitude and longitude—but many of today's most complex GPS applications also require a standardized vertical, a common baseline of "zero elevation." This is especially true for climate change research, which must account for millimeter-level changes in sea level.

One of the most powerful tools enabling these precise vertical measurements comes from the two satellites that form the foundation of the Gravity Recovery and Climate Experiment. GRACE, a project run jointly by JPL and Germany's national space agency, offers another example of the interdependent feedback-loop quality of a world heavily dependent on GPS, in that it *requires* GPS to perform its duties but also *enables* GPS to perform at the highest level. The GRACE satellites orbit the earth at an altitude of 300 miles, 140 miles apart. The satellites continuously bounce a signal off each other, measuring the time it takes the signal to travel to the other satellite and back. The clocks on both satellites remain synchronized by taking their time signal from GPS.

The variations in the time it takes the GRACE signal to make the round-trip journey translate into small deviations in the 140-mile gap between the satellites. These in turn reveal small variations in Earth's gravitational field. The amount of force gravity exerts on objects is influenced by topography—the presence of mountains, for example, increases it. But it is also affected by the ongoing and constant redistribution of groundwater above and below ground, which shifts the planet's center of mass. Much like sound waves, gravity itself decomposes into harmonics. The GRACE satellites measure these high-level gravitational harmonics. If the lead satellite pulls ahead of the other—the system can detect a change as small as a thousandth of a millimeter—it is experiencing an increase in the gravitational field.

Like any other object in the earth's gravitational pull, GPS satellites are affected by the location of the planet's center of mass. The changes in the gravity field affect their behavior, so GRACE ultimately enables precision GPS receivers—the kind used to measure plate movement, earthquakes, volcanoes, and provide surveyors and scientists with exact distant measurements—to base their calculations on an absolutely correct understanding of the satellites' orbits. "It's an infrastructure mission that makes GPS better," JPL's Bar-Sever says, "and all other science cascades from that."

From 2007 to 2009, a network of fifty GPS stations was set up around the perimeter of Greenland's ice sheet, the second largest in the world. The ice, thousands of feet thick, lies on top of bedrock, pressing down on it. Like a bathroom scale that measures a body's weight based on how much it can compress a spring, the GPS network measures the melting of the ice sheet by noting the elastic effects of the bedrock, which "bounces back" in response to the decrease in ice. The uplifting of the bedrock is reflected in the changing vertical position of the GPS stations. During 2010, some of the stations rose by nearly a quarter inch, the effect of an unusually large number of days warm enough to melt the ice. The distribution of this "uplift anomaly," detected by GPS, correlated with a similar unusually large deterioration of the ice sheet—a "mass anomaly"—detected by GRACE data, with the stations experiencing the greatest uplift matching the areas experiencing the most melting ice.

What might Alfred Wegener have made of all this high-tech activity, enacted to apprehend unfathomably complex atmospheric and meteorological processes—especially in his beloved Greenland? Still unable to persuade many people of his continental drift theory, Wegener returned to Greenland for a fourth time in 1930, leading a group of scientists. To test certain meteorological theories, and to measure the thickness of Greenland's ice, Wegener wanted to establish an observation station in the middle of the ice sheet along the 71st parallel, the same latitude as existing stations on both coasts, 250 miles away. The location for this "Mid-Ice" (*Eismitte*) station was at an altitude of 10,000 feet. To reach the point required was a massive logistical and physical undertaking, made more difficult by icy conditions that prevented the expedition's ship from landing when it arrived in April. Wegener and his team did not begin establishing the western station until the middle of June. A month later, Johannes Georgi and Ernst Sorge, a glaciologist, left to build the Mid-Ice camp.

With winter rapidly approaching, Wegener set off in September with another meteorologist, Fritz Lowe, and thirteen Greenlanders, to resupply Mid-Ice. Conditions made the already difficult journey up the ice sheet excruciating. All but one of the Greenlanders soon turned back—incredulous at their comparative lack of zeal, Wegener "could hardly get over this," Georgi recalled—but Wegener, Lowe, and a twenty-two-year-old named Rasmus Villumsen struggled on, through snow so fine that the sleds sank into it. The trek, which should have taken two weeks, lasted forty days. The men finally arrived at the end of October. Lowe was too weak to make the trip back immediately, with badly frostbitten hands and feet. (He would lose his toes throughout that winter, his colleagues at Mid-Ice amputating them with scissors.) On November 3, two days after everyone celebrated Wegener's fiftieth birthday, he and Villumsen began the return trip to the western station.

Sometime during the return journey, the exertion became too much for the continental drift theorist. Wegener died in his tent of an apparent heart attack. Villumsen carefully buried Wegener's body in a sleeping bag and a reindeer skin, retrieved Wegener's diary, and embarked once again for the coast. He was never seen again. Wegener's colleagues found Wegener's body six months later. They built an ice-block mausoleum over it and later marked the spot with a 20-foot iron cross.

The Greenland GPS stations near the 71st parallel exhibited

some of the largest uplift anomalies in 2010. After nearly ninety years of winters, Wegener lies under at least 300 feet of ice. According to some calculations, the glacier's drift, accelerated by climate change, will eventually deposit his body on Greenland's west coast.

CHAPTER NINE

Tied Together (40.74375° N 73.9835° W)

In the photograph, a body lies in the street, surrounded by people. The scene is perfectly framed, the perspective slightly elevated above the crowd. The photographer was not part of the group: just an observer. The angle gives the photo a distinctly voyeuristic quality. There is no other information known about this photo. No context, no photographer, no caption, no source—just an image depicting a body prone on the pavement. The body appears to be male. Something bad is happening to this man—

© 2015 Google

nobody voluntarily lies down in the street. One way or another, this man has fallen. I've taken to calling him "Fallen Man."

Beyond these observations, we are in the realm of pure supposition. Is Fallen Man having a heart attack? A stroke? Is he already dead? Maybe he just fainted. Maybe he's drunk. The faces of the onlookers are blurred, but their body language gives us some clues. The woman clasping her hands to her head, or the one behind her holding her hand to her heart—these gestures suggest shock or concern. But the man with the shoulder bag—why is he standing so close to the body without appearing to act? (But then, how could he? His legs are severed below the knees.)

There is one thing I know for sure about this photo: the location it depicts, the northeast corner of East 29th Street and Park Avenue South, in the Kips Bay section of the New York City borough of Manhattan. I know this because the photo is labeled as such, but I also recognize it. That doorway where a man appears to be expressing concern is the entrance to Park Avenue Audio, a store that sells high-end audio and video components.

I first saw the photo in 2009. Fallen Man was one of many strange images uncovered by people exploring the millions of 360-degree ground-level photos that comprise Google Maps's Street View archive, which contains millions of searchable 360-degree geotagged ground-level images, captured by fleets of cars with special rooftop-mounted cameras. Through tedious combing or pure luck, anonymous Street View spelunkers spotted it and circulated it online. It generated some small, inconclusive pieces in a few media outlets, but then Fallen Man disappeared. By the time I saw the image, Google had replaced the photo with another photo of the same scene. The Fallen Man had fallen off the face of the earth.

Among the many strange panoramic images Google's cameras have managed to catch—house fires, home invasions, roadside urinations, holdups, flashers (word gets around when a Street View car is making its silent canvass in an area)—Fallen Man is among the starkest documentary captures. Like the others it is stripped of any contextual information. These photos carry an implied intimacy, but they never tell you exactly what you are seeing, and many of them probably don't show what you think they show.

Looking at Fallen Man's image is like doing a jigsaw puzzle in reverse. The scene is complete, but we don't know how the pieces fit together—and the more we try to carve it up, reduce it to its component parts, the more the pieces crumble into pulp. The man in the pink shirt standing in the door is Glenn Marston, a store employee. His pose reads as purposeful, as though he were striding out to take in the commotion. But Marston, who no longer works at Park Avenue Audio, has only a vague recollection of the incident. He has a much more vivid memory of the time a peregrine falcon landed on that corner—or the time that he aided another fallen man, who had suffered a seizure in that very same crosswalk. "I grabbed some OJ and gave it to him," Marston recalls. "He quickly revived, then walked away without even saying a word."

Another bystander, Tyler Barnet, has a vivid memory of coming upon the scene of the Fallen Man. "I looked at the guy's face," he says. "He looked Indian, and he looked dead—and I said that to the man I was walking with." The man was a business associate. "We were discussing technology," he recalls. "He made a remark that the first iPhone was coming out next week, and was going to change everything. When I walked him back to Penn Station from 31st and Park, we saw the body and he became weird. I am no longer in touch with him."

The first-generation iPhones went on sale in June 2007—and the light clothing worn by the people surrounding the Fallen Man suggests a spring or summer tableau. Now we're getting somewhere. We appear to be able to triangulate the date when the image was captured. But there is a problem. The scene, apparently unbeknownst to Google, still resides in the Street

View database. You won't see it if you try to approach the corner by creeping down 29th or Park—but if you drop in via one of the posted photos of Park Avenue Audio at the bottom of the screen, the Fallen Man is there. And the time stamp on the photo places it in 2009.

"That is the scene I remember!" Barnet wrote in an email after I sent him these phantom images, which I didn't discover until a few years after I saw the original Fallen Man photo. "I saw it when I lived in a certain apartment, and that was only for nine months after I graduated college in 2006. I remember him looking just like that—and it was on I was living on East 31st Street." Who's to say his memory is any faultier than Google's time stamp? Google's efforts to wipe the image were obviously haphazard—and it's worth noting that visible time stamps did not begin appearing on Street View until several years after 2009.

Any way you look at the Fallen Man—as a photograph conveying a message, as a conglomeration of data that coalesces into tangible information—the more the image fades. We will probably never know who the Fallen Man was, when he fell, or why. All we can cling to is our knowledge—solid, dependable, based on my own familiarity with the city I called home and my own real memories of that corner as a part of my cognitive map, in other words, my *perfect knowledge*—of where he fell.

If all other attempts to understand the meaning of Fallen Man have failed, why should the knowledge of his location be any less contingent? GPS gives us the illusion of infallibility; the photograph is geotagged to specific GPS coordinates. Those coordinates are linked to latitude and longitude and other systems that we use to define physical space. But the deeper you drill into the data, the more you realize the fragility of this global system we've constructed, and upon which so much of our society now relies. In an age when GPS gives us a blue dot on a map—and perhaps also a rich visual image to go along with it—it becomes increasingly difficult to understand that this system is imaginary.

The data appears solid, but it is only solid in relation to something else. We know a great deal about where Fallen Man fell, but at the same time we know almost nothing at all.

Fallen Man, Street View, and Google Maps would not exist without GPS, but they are really made possible because of the dramatic collusion between GPS and geographic information systems (GIS). Unlike GPS, GIS is not a specific technology, but rather the creative merger of computer science, cartography, and database management. The rise in popularity of GIS in recent years, especially as so much of the world has become data-obsessed, has been just as dramatic as the growth of GPS use. Esri, a California-based GIS software company launched in 1969 by Jack Dangermond, along with his wife, has allowed him to amass a net worth of $3.1 billion, putting him among the 300 wealthiest Americans.

For most of us who use GPS, what we are really using is GIS. Those GPS coordinates do not mean much to us without a map on which to peg them. A historical and technological survey of GPS published in 2002 summarized the relationship between the two technologies/concepts: "GPS receivers without GIS have no knowledge of the real world."

This knowledge runs deep. Street View ultimately taught me nothing definitive about Fallen Man, but Street View images are perhaps thicker with real information than we realize. An MIT Media Lab study of "perceptual inequality" showed experimental subjects pairs of images selected randomly from a set of hundreds of street-level images of New York City, Boston, and two cities in Austria, Salzburg and Linz. For each pair, subjects answered the questions "Which place looks safer?" and "Which place looks more upper-class?" The study found that both positive and negative impressions were more clustered around geographic regions in the American cities, and more widely dispersed in the Austrian cities, suggesting that the cityscapes of Boston and New York City exhibit starker inequality. To test the soundness of their experimental methods, the researchers compared the New York clusters with neighborhood homicide rates, and found a strong correlation. Street View's view of the world can apparently teach us a lot about it.

But to say that GIS helps a GPS receiver know the *real* world—knowledge it then imparts to us, the users—is not quite right. Like any map, the GIS component of GPS offers an actively mediated representation of the world. GPS therefore constructs its own version of the world, not a mirror image of "reality." To represent geographic data, Google Maps uses the Mercator projection, named after the sixteenth-century cartographer Gerardus Mercator, which represents latitude and longitude as being perfectly perpendicular. Sailors and navigators liked this projection because a ship on a steady bearing was represented by a straight line on the map. As any user of *etak* (the ancient Pacific navigational aid, not the 1980's proto-GPS car guidance system) will tell you, this is a view of the world completely at odds with the way parallels and meridians behave on the earth's curved surface, and grossly distorts the size of land masses the closer they are to the poles. The Mercator projection remained popular throughout the centuries. In the 1980s, cartographical societies drafted a resolution that discouraged any further use of it, in favor of equal-area maps, which more accurately depict the relative size of land masses.

Around the same time, the first mass-market computerized maps appeared. Some used the Mercator projection, others did not, and well into the 1990s there was no clear preference. For Google Maps, which appeared in 2005 and soon became the world's most accessible GIS, the Mercator projection was an obvious choice. At every point on a Mercator map, "north" points in exactly the same direction, making it more suitable for panning and zooming. Although a Mercator map distorts large land masses, it accurately portrays more localized geography—the opposite of what occurs with equal-area maps. Since Google Maps is so often used as a local reference source—where is that restaurant, again?—local accuracy is more important than global accuracy.

Something we don't usually notice—but which becomes apparent if we zoom out as far as possible—is that Google Maps is giving us the world in an aggressively distorted form. Google tweaks our perception by using a version of the classic projection called Web Mercator, which is identical to its forebear save for one crucial difference. GPS receivers translate GPS data into latitude and longitude points using, as reference, an irregularly shaped model of the world called an ellipsoid, which closely approximates the actual shape of the earth. Web Mercator considers the earth to be a perfect sphere. Google Maps receives latitude and longitude computed using one model of the planet, and then does its own computation to project it onto a visual interface that uses a different model. If you attempt to use Google Maps to plot a constant bearing over a very long distance, this discrepancy will become apparent.

Why introduce this distortion? Because performing computations on a sphere is much simpler and faster than performing them on an irregular shape. Daniel Strebe, a software engineer and expert in map projections, estimated that if the hundreds of millions of computer maps rendered daily by map-based services such as Google were projected using the ellipsoidal Mercator, the annual cost of the increased electricity usage would be tens of millions of dollars—and the energy required to generate it would have a measurable environmental impact.

This is one way to express the looming presence of GIS. Not only does it shape our perception of the real world—the decisions we make regarding how we use it can have actual acute physical effects on that world. GIS has brought me into close contact with the Fallen Man, tricked me into thinking that I was party to privileged knowledge about him and what he had experienced. But I knew nothing about him—a true fact underlined by the meaningless conglomeration of maybe-facts that I had gathered. I would never know anything more about him, except for where he fell. Behind that knowledge was a whole series of processes, a vast technological apparatus that told me where I could find him. I needed to know more, and I thought I should start with how Street View comes together.

Google would not respond to my requests to ride around in one of their data-collection vehicles. TomTom was more accommodating. The Amsterdam-based company is one of the GPS industry's major players, roughly to Europe what Garmin is to North America. After launching in 1998 with a mobile phone mapping program, TomTom released its first GPS navigation system in 2002. Like Garmin, TomTom's products did not take off until they shed all vestiges of peripherals, such CD-ROMs from which maps had to be uploaded. Unlike Garmin's Gary Burrell and Min Kao, TomTom cofounder and CEO Harold Goddijn had a business and finance background, but he knew what he needed from a GPS unit: "buy, take out of box, drive home." By the early 2010s, TomTom accounted for just under half of all consumer satellite navigation systems in Europe, and Goddijn and his wife, Corinne Goddijn-Vigreux, with whom he founded the company, were among the 100 wealthiest people living in the U.K.

It was around this time that TomTom, which had about 21 percent market share in the U.S., agreed to let me ride around in one of four TomTom data-collection vans currently operating in North America. The company was trying to increase its U.S. presence, an effort that included updating its American maps. TomTom had recently outbid Garmin to acquire Tele Atlas, one of the world's oldest and largest geographic data collecting companies. I met up with an affable recent college grad named Joe Palatucci, who was driving an unmistakable green van emblazoned with TomTom's logo and filled inside with computers, components, and wires. Palatucci was living an itinerant existence, staying in motels around the eastern seaboard, not bothering to keep a fixed address.

Palatucci explained that TomTom was in the midst of trying to account for new highway construction that had arisen from the 2009 federal stimulus package. Today's job was to drive the major highways and freeways of western Long Island and eastern Queens, near JFK airport. The sight of the van, whose purpose is immediately clear, tends to elicit some odd responses, Palatucci said. He'd seen a lot of rude gestures lobbed his way, been flashed at least once, and had a drunk middle-aged woman in Rhode Island fling a bottle at the van. "In Vermont, for some reason, they'll hang their body out the window and go 'Whoo!'—which freaks me out because I'm afraid they're gonna fall in front of me." New York, though, was more sedate—"people here aren't impressed by a lot of things."

As he drove, he kept one eye on a monitor to his right. A laptop computer was connected to two servers in the back of the van. A camera on the roof snapped photos at a rate of 2.6 per second, while five laser-pointing devices measured the distance from the van to passing signage. Along with GPS navigation, the system was also running an inertial navigation, so that if the GPS momentarily cut out—a not uncommon occurrence when going under a bridge—the location data would not falter. System hiccups still occurred. Every 10 meters, a switch told the camera to take a picture, but bumps often jarred the odometer, which then failed to trip the switch, necessitating a second trip over the same stretch. Joe also had to be attentive to weather conditions, adjusting the camera for different light conditions. After driving for about two hours, we'd amassed 100 gigabytes of data.

At the end of the day, Patalucci downloaded the data and sent it to TomTom. It was analyzed and compared to existing map data, and the current layout was corrected accordingly. Much of that work was automated. The system could spot what looked like anomalies—cars driving in places where the previous system showed no roads—and it could use the laser data to recognize common signage. But large swaths of the data were still pored over by humans in production departments in India and Poland.

Today, the process would be even more automated. One of TomTom's strategies for dealing with a world of GPS-enabled smartphones is to promote its traffic data service. Millions of TomTom users allow the company to plot their location data. By crunching the numbers, TomTom learns about a region's traffic patterns and congested routes. Even if a user is running TomTom software on a phone, accelerometer data can place that phone in a moving car. All this data makes it easier to spot the existence of new roads and the disappearance of old ones, and obviates the need for constant data collection. Vans will only be sent out if an analysis points to many changes in an area.

Because of TomTom's international presence, it maintains map data in several countries. A few years after my day in the van, the company was nearing the end of a project to use all that mobile mapping to realign its maps. Hervé Clauss, TomTom's director of global imagery and sourcing, told me this is a challenge because every country has its own "projection methods," its own way of defining distance and positioning, making it difficult to square all the data into one database. "If we had to convert in each and every country, it would be a painful exercise," he explained. "We need to use one reference system. So we use WGS 84."

I recognized the name, because I knew WGS 84 had some foundational relationship to GPS, though the meaning of the connection eluded me. If TomTom was plotting GPS data and putting them onto maps, and GPS was universal, what was this other thing that was required? "If you said you wanted to put a point on a piece of paper, on the coordinates 10-10, x and y coordinates, you'd need to set up your grid on a piece of paper, because that could mean ten inches or ten centimeters," Clauss explained. WGS 84 was that grid. "That's basically where it all starts," he said. "All the tools we have in the company use the same reference numbers. Throughout the company, we talk the same language."

I had made the mistake of assuming that GPS was this language. GIS, in the form of Google Maps Street View, had allowed me to visualize the place where the Fallen Man fell, as a spot near the northeast corner of a particular New York City intersection. It had done so by translating GPS information into the language of latitude and longitude. But what was the link between GPS and these lat/long coordinates? GPS had to have a model of the world within itself, to access when making its computations. Knowledge of where the Man fell to Earth—and, presumably, every other GPS location—was somehow a function of WGS 84. So now I needed to understand this common language, this mediating grid.

The modern science of geodesy—refining our ability to measure the size of the earth and its gravity field—extends back more than 2,000 years. The Greek astronomer Eratosthenes used observation of the sun and the measured distance between Alexandria and the Egyptian town of Syene (now Aswan) to compute the planet's circumference. His conclusion—25,000 miles—is only about 100 miles off from the figure we use today.

The related practice of land surveying—attempts to get an accurate sense of spatial relations on the planet—is even older, traceable to ancient Egypt. The first land survey to use modern methods—essentially, the first organized attempt to define "here" and "there" over large swaths of land—was undertaken by the French astronomer Jean Picard in 1669, and continued by the Italian astronomer and mathematician Giovanni Cassini after Picard's death. To fix locations, they used triangulation. The concept is simple geometry. Imagine the three points of a triangle. You know the distance between point A and point B, the angular heading of the line connecting point A to point C, and the angular heading of the line connecting point B to point C. You can use this information to determine the lengths of the two unknown sides of the triangle. One side of the triangle then forms the basis for another triangle. That triangle spawns a third. And so on, triangles moving across the earth, all ultimately referenced to that initial baseline, point A to point B. This is called an arc of triangulation.

It took a special kind of obsessive mania to actually make these triangles multiply without losing cohesion. One small error would ruin the whole survey. Picard began by using wooden rods to measure a 14-kilometer baseline that stretched from Paris to Fontainebleau. The measurements eventually extended south to the Spanish border in the Pyrenees, and north to Dunkirk, on the English Channel. Picard's two arcs, one moving south and the other moving north, revealed something strange about the planet. The surveyors had used their measurements to determine the size of one degree of latitude, but the northern arc and the southern arc yielded different figures. The earth, it turned out, was not round.

Picard thought the planet was shaped like an upturned egg, with a polar axis longer than its equatorial axis. The Dutch mathematician Christiaan Huygens, building off work done by Isaac Newton, disagreed. He thought the equatorial axis was longer, making the planet look more like a squat grapefruit. After years of debate between the two camps (egg and grapefruit), Huygens's theory was proved correct. The geometrically correct term for the earth's grapefruit shape is "oblate spheroid." Any geodetic observation, any land survey, begins with a reckoning of the precise shape of this grapefruit. The mathematical representation of this shape is called the reference ellipsoid.

"Prior to the 1950s, geodesy was a very mathematically theoretical field," says Thomas Herring, the MIT geophysicist. "And the reason was—conceptually, anyway—we were trying to measure the whole shape of the planet. The work of some of the great mathematicians was, in fact, motivated by wanting to be able to determine the shape of the earth, while you could only see little pieces of it at any one time." As geodesy diversified, this global concept splintered, becoming more localized and more practically oriented. For governments to calculate land taxes, they needed thorough and accurate surveys of their region, to determine exactly how much space a landowner possessed.

The early land surveyor faced a dilemma related to the infuriating subjectivity of all location data. The position of any spot on earth—its latitude and longitude—can be determined using astronomical observations. Alternately, it can be defined geodetically, via baselines and arcs of triangulation. In theory, careful use of either method should yield the same values, but in reality they do not, because of "the little kinks and bends in the gravity field," Herring says. A surveyor using astronomy is not relying on an ellipsoid, but on something weirder and more abstract called a geoid. The geoid is an orderly imaginary earth where the force of gravity, anywhere and everywhere, is exactly perpendicular to the planet's surface. Let a coin fall from your hand while standing on the geoid, and it travels along a line perfectly perpendicular to the ground. Because of disparities in gravity on our actual planet, the geoid is a convenient fiction, one that flummoxes any "nonfiction"—that is, "real"—attempts at fixing a position. Every triangulation calculation requires a calculation of how the gravity field at that part of the earth tweaks how the coin falls from the hand. This "ellipsoidal vector" is the angle that best describes the pull of gravity.

So the early land surveyor had to cook the books a little. To survey a large piece of land accurately required triangulation—which in turn necessitated some way to make those measurements conform, as much as possible, to those gathered by astronomical methods. "So you'd look at the astronomical latitudes and longitudes, and your geodetic latitudes and longitudes, and you'd work out the best shape of an ellipsoid you could use to make those things as close as you could to the same values," Herring says. "And so different countries came up with different ellipsoids and different orientations."

An ideal ellipsoid is one that does the best job of reconciling those values. It is a shape that most efficiently aids local surveyors. The popular Clarke Ellipsoid, for example, developed by the British geodesist Alexander Ross Clarke in 1866, defines the earth's equatorial radius as 6,378,206.4 meters long, and describes its grapefruit shape by using a flattening ratio—the difference between the lengths of the equatorial and polar radii—of 1/295.

And yet . . . the surveyor still needed more than an ellipsoid. The surveyor needed to stretch a grid across that ellipsoid. We call this grid a datum. The datum makes the Cartesian coordinates possible. It is usually achieved by picking a spot, a kind of ground zero of surveying, and defining every other spot in relation to it.

You can begin to see the inherent problems in any attempt to spatially define the earth as a unified monolith. Armed with an ellipsoid and a datum, the surveyor surveyed. Across the world, other surveyors were doing the same, often using datums that were relevant only within the borders of one country. The system worked on a local level, but it made the world a patchwork of measurements. In the lingo of geodesists, the world was not "tied together."

The modern history of surveying in the United States begins with Thomas Jefferson. In 1807, while serving as the nation's third president, he established the Survey of the Coast, America's first science agency, run by a Swiss expatriate named Ferdinand Hassler. The actual surveying work began in 1816, with a survey of New York Harbor. (America's first official "here" was a few miles south of Fallen Man's corner.) Hassler began by measuring two baselines, one along the shoreline at Gravesend Bay (the future location of Coney Island), and one near Elizabeth, New Jersey. By the following year, he had created a small triangulation network and eleven observation stations.

Triangulation was a slow, painstaking process, involving a theodolite—a telescopic device that measures angle—and chains to measure distance. By the time of Hassler's death, in 1843, after he suffered serious injuries while trying to protect his equipment during a storm, the arc of triangulation he and his team created only extended to southwest New Jersey, near the shore of the Delaware River. It would eventually evolve into the Eastern Oblique Arc, completed in 1898, stretching from Maine, near its border with New Brunswick, down to the southern edge of New Orleans. Around the same time, the Transcontinental Arc was completed, connecting Cape May, New Jersey, to a lighthouse near Point Arena in Northern California. In the area between Colorado and the Pacific, some of the triangulation lines were over 100 miles long, with surveyors hauling materials and assembling observation towers in barely accessible locations.

The large arcs were facilitated by the adoption of the New England Datum, America's first, in 1879, based on the Clarke Ellipsoid and tied to an observation station near Chesapeake Bay. In 1901, it was renamed the United States Standard Datum, with a control point at Meades Ranch, 150 miles north of Wichita, Kansas, near the geographic center of the country. When Canada and Mexico agreed to reposition their own national triangulation surveys, tying them to Meades Ranch, it became the North American Datum of 1927, or NAD27.

By the start of World War II, every advanced nation in the world had its own internal surveys and datums, some created by tying together existing local surveys, and others the result of extensive national surveys. When the U.S. Army realized that confusion over the various European datums was causing its troops to miss artillery targets, it turned to a geodesist for help.

Fred Hough had worked for oil companies and the government for twenty years when he was persuaded to join the war effort overseas. When the American infantry entered a German city, the "Hough Team"—Fred and his handpicked crew—were the first behind them. They would plunder map and surveying information from libraries and government institutions, and later use the information to update the U.S. military grid.

The Hough Team's big score came in the spring of 1945. In Saalfeld, a village in central Germany, they discovered several boxes of materials relating to Russian geography and geodesy captured by the Germans. That day, Saalfeld was about to be turned over to advancing Soviet troops. Hough assumed they would want this material, so he loaded up several artillery trucks. The last of them was hauling the treasure to the American zone just as the Soviets entered the town. Hough's team combed through the information, piecing it together with geodetic data gathered by other European countries, based on their own national datums, eventually using thousands of different maps to create the first standardized military map of Europe. It went on to be a primary tool in the creation of the European Datum 1950, which served the same unifying purpose for Europe as NAD27 did for North America.

The geopolitical significance of geodesy increased with the onset of the Cold War. In a very real sense, the West did not know the exact location of the USSR. "Missiles were the big drivers in getting the datums tied down," Gaylord Green remembers of the days when he was helping to build GPS as a member of Brad Parkinson's team. "If I wanted to hit a target in Russia, I couldn't hit squat if I didn't have their datum tied down to mine." The geodetic problem was bigger than merely conflicting coordinates. The trajectory a missile follows is influenced by the gravity field where it is launched, and its aim can be disrupted by the gravity at the target, so countries often kept their gravity data classified.

By the early 1950s, geodesists at Ohio State University's Mapping and Charting Research Laboratory, much of their work funded by the Air Force, were leading an intense effort to tie the world together. They measured gravity fields and experimented with photogrammetry (using images to extrapolate location information). Geodesists adapted the radar concept, perfected during the war, to pioneer electronic distance measurement (EDM), sending and receiving signals between two points to measure distance. EDM was the biggest breakthrough in surveying since triangulation. It obviated the need for a line of sight between surveyed points—a necessity with theodolites, a common surveying tool—with results accurate to within centimeters.

The Ohio State group did the first major surveys of isolated Pacific islands and archipelagos, including linking the Marshall Islands weapons testing zone to NAD27. (While preparing to attack the Marshalls during the war, the Navy discovered that the archipelago was about 10 miles from where the charts said it would be.) The Army Corps of Engineers sent teams around the world to tie down loose geodetic ends. Around the same time as Sputnik, one group surveyed a remote section of the 30th meridian, which runs from the tip of Norway to South Africa. They dodged wild animals, erected and dismantled 100-foot-high observation towers, tall enough to rise above the jungle's tree cover, and worked through the night, aiming the beams of their theodolites at one another to measure angles.

An accurate assessment of the 30th meridian would refine our measurement of the earth's circumference, *Life* magazine noted—"just the kind of geodetic fact which may win or lose a war" if the intercontinental ballistic missiles were ever launched. (For every three American geodesists, the magazine ominously noted, the Soviets had 500 of their own.) But it was also widely known among scientists that the promises of the space age made our lack of geodetic knowledge all the more glaring. "It's ridiculous not to know in 1958 exactly what the distance is from New York to Moscow or Mandalay or anywhere else," one scientist complained. Like many of his colleagues, he hoped work done during the International Geophysical Year would refine these measurements.

As much as the International Geophysical Year activities during 1957 and 1958 were focused on outer space, this question of distances preoccupied a sizable contingent of participants. The secretary of the committee overseeing Great Britain's contributions to the effort noted that the U.S. Air Force had only recently determined the exact distance between Washington and the coast of France—and kept the information classified. "This is not so silly as it sounds," he said, "as apparently uncertainties exist in intercontinental distances to as much as 300 feet." John O'Keefe, chair of the geodetic division of the U.S. Army Map Service, summed up the problem of gazing at the sky so much we forgot about the ground beneath our feet. "For all our talk of exploring outer space, we still have only the haziest idea of where we are on earth," he said. "Our latitudes and longitudes are fuzzy."

The project to build an artificial satellite, the global endeavor's grandest project, also had geodetic aspirations. In the original proposal drafted by the Naval Research Laboratory for Project Vanguard, the satellite-development project that was upstaged by Sputnik, geodesy is the first subject discussed by the authors, GPS visionary Roger Easton and Milt Rosen: "The United States has been and is now expending considerable effort and funds to refine geodetic measurements. One of the most difficult geodetic tasks involves tying together the various continental grids and locating the many islands with respect to these grids. Whereas it is possible to map entire continents by well-known methods of triangulation, the technique fails when it must be extended across large bodies of water. As a result, geodesists have turned to other methods which are, in general, based on the use of the only available satellite, the moon, as a measuring vehicle. . . . An artificial satellite, which would be small in size and mass, and at a short distance from the earth, would have numerous advantages over the moon."

This promised to be the next big leap forward for geodesy, after triangulation and EDM. "With artificial satellites, you had the potential to look up from the ground and see the satellite from widely separated places," Herring explains. "And that allowed you to start connecting those locations together."

But the idea of satellites as the ultimate arbiters of location carried a maddening paradox. Latitude and longitude are ultimately referenced to the center of the earth. But we can't determine the exact location of that center without knowing the exact size of the earth. One could discern the shape and volume of the planet by apprehending a satellite from various locations on earth. But for *that* to work, one needed to know the exact position of the satellite at any given moment, and that position could only be determined with absolute certainty if the satellite was referenced to . . . the center of the earth. A few years before Joseph Heller's *Catch-22* provided the locution for this geodetic conundrum, here was a space-age example of the concept: for satellites to help us locate the planet's center, we needed to know the location of the planet's center.

Pinpointing the earth's center offered the promise of a truly objective geodetic reference system. Unlike the world's many datums, doomed to never quite align, an earth-centered reference system would have no need for a control point anchored to a single spot on the planet's surface. As the science of plate tectonics was making increasingly evident, any such point was always in flux, making it an untrustworthy anchor. And so scientists turned their energies and attention to the earth's core. The Pentagon's Defense Mapping Agency arrived at the first iteration of the World Geodetic System in 1960, arguably the most underappreciated scientific achievement that grew out of Cold War imperatives.

Once the World Geodetic System was established, it was tightened and refined. By the time of the Williamstown meeting in 1969, at which scientists discussed the need for better tools for measuring the earth, the position of Europe relative to North America was known to within about 50 meters, roughly half of what the margin of error had been ten years earlier.

From its inception, GPS used the World Geodetic System for its datum. In the early 1980s, the Defense Mapping Agency embarked on a massive revision of WGS. The new version, WGS 84, was accurate enough for the current generation of nuclear missiles, and also tight enough to predict the orbits of the GPS satellites with extreme precision. The availability of GPS, and the refinement of its accuracy, brought the satellite revolution to land surveyors, the second industry (after precision timing) to embrace GPS. "Surveying really had not progressed a lot since the time of George Washington," Charlie Trimble says. "The only difference is that we had laser distance measuring, so you didn't have to use an actual chain. Fundamentally, what GPS did is it replaced the requirement for a line of sight between a known point and an unknown point on earth. So it made virtually every surveying job take the same amount of time, no matter whether the distance between the known and unknown was a block or 20 miles."

Since its debut, WGS 84 has undergone periodic readjustments to make it accurately reflect the planet as it is. This earth-centered datum has succeeded in tying the world together—especially as the primary instrument of its expression, GPS, is such a universal, globally available utility. In parts of the developing world, governments are using GPS for some of the same political reasons that drove the use of triangulation in the early days of modern surveying. Until recently, official maps in Burkina Faso were based on data gathered in the early twentieth century, using a network of unstable astronomical reference points. In the country's immediate postcolonial period, French and American mapping officials used triangulation to tie the country to the Blue Nile Datum. The country was later resurveyed based on the WGS, and in 2012 the government established a series of continuously operating GPS receivers linked to the International Terrestrial Reference Frame (ITRF), the ultra-accurate frame used by geophysicists to monitor plate movement. The Burkina Faso government's embrace of this highest of high-tech physical reckoning is an attempt to modernize a land tenure and management system rife with instability and inaccuracy. A World Bank report predicts that the system will help "avoid land ownership overlapping . . . and enhance social equity and peace."

The ITRF is considered the ultimate mathematic representation of the earth: its size and the exact location of its center. It is acknowledged as the ultimate datum. WGS 84 is the United States Department of Defense's "realization" of that frame, an attempt to wrap a grid around the skeleton as tightly as possible, so that, today, WGS 84 is accurate to the ITRF to within a centimeter. Who controls WGS 84? The same people who created the first iteration of the World Geodetic System in 1960. It's no longer called the Defense Mapping Agency. Reflecting its current priorities, it is now the National Geospatial-Intelligence Agency (NGA).

One of the more obscure outposts of America's sprawling intelligence infrastructure—even President Obama, when he met an NGA employee during a 2009 photo op, appeared unfamiliar with the office and its purview—NGA was briefly in the spotlight in 2011, for its role in the mission that killed Osama bin Laden. The Department of Defense designates NGA as one of four "combat support agencies" that provide tactical support for U.S. military maneuvers. "We have a global mission to map anywhere on the planet," says Steve Malys, an NGA geodesist.

To provide the most accurate maps in the world, NGA needs a perfect model of the world. Just as it has from the beginning,

GPS and its continual refinement reflects military imperatives, unknown to most GPS users but always there in the background. By controlling WGS 84, NGA is essentially the final authority on the GPS data used by everyone—including, in a one-world irony—America's enemies. This was the ending point for my quest to find the source of Fallen Man's location.

"We know the center of mass of the earth to within less than the size of a postage stamp," says Steve Malys, an NGA geodesist. "That's the starting point for all geopositioning. For our purposes, we're satisfied with one centimeter. There are recommendations from the National Academy of Science for GPS to go toward one millimeter, but we're not there yet."

NGA is arguably as important as the Air Force in making the gears of GPS turn. Of the seventeen monitoring stations around the world that track the progress of the GPS satellites and feed the information to the GPS Master Control Station in Colorado, eleven are controlled by NGA. They most likely serve other functions related to NGA's intelligence gathering. A very prominent and respected geophysicist told me that NGA's monitoring station in Australia, near Adelaide, contains a room only accessible to American citizens, a fact Malys will neither confirm nor explicitly deny.

NGA's cutting-edge gravimetric research has helped its geodesists produce the most precise breakdown of the earth's gravity field. "It represents the complicated potato shape of the earth," Schmidt explains, shattering my image of the planet as a galactic grapefruit. "To do that required 4.7 million coefficients." As NGA's representation of Earth approaches (but will never quite reach) the complexity of the geoid, it occasionally recommends tweaks to WGS 84. NGA forwards them to the GPS Master Control Station, which integrates the new data into the satellites' software.

When this happens, unbeknownst to the rest of us, our image of the world changes, just a little bit. Although we will never notice the difference, it is this constant refinement that gives us a more perfect understanding of the ground beneath our feet, pinpointing the respective spots where we will all eventually fall.

The 1969 Williamstown Report, the NASA-sponsored manifesto that called for a concerted effort to improve measurements of the earth's physical properties, made an important point regarding the subjective nature of all locations. Nothing is truly fixed. Like a spinning top, the planet wobbles as it spins, the gravitational pull of the sun and the moon changing the orientation of the earth's rotational axis in cycles that last about 26,000 years. At any given moment, "the principal axis of inertia," the line going straight through the earth's center, is not necessarily the axis line around which the world turns. Even if it did, so what? It's not like this principal axis is fixed relative to some unmovable, unchanging set of axis points, because the distribution of water around the planet changes the location of its center. Nothing is where it seems, because it is nowhere.

For scientists, the report noted, one problem associated with the wobble (the technical term is precession) is that it is difficult to differentiate from the "galactic motion of the stars." Nowhere, it seemed, was there any rigid, unmovable beacon, something stable around which to base a measurement system. Nowhere nearby, anyway. We could, however, use quasars—distant galaxies centered around huge black holes—which are far enough away (millions, even billions of light years) that they appear, from our perspective, to be fixed points, and are also bright enough (imagine the combined luminosity of the few hundred billion stars in the Milky Way multiplied by 100) for us to know they're out there. This is why we needed VLBI, the report argued. If we could measure precession more accurately, we could more accurately predict the motion of quasars in relation to Earth. And if we could establish VLBI beacons on other planets in our solar system, we could "fix the position of the ecliptic"—the path, seen against a background of stars, that the sun appears to take from the perspective of Earth.

VLBI revolutionized plate tectonics research, until GPS offered a similar but less expensive and more mobile alternative. Still, both methods remain earthbound. "We don't have GPS on Mars," says Tomas Martin-Mur, an engineer at NASA's Jet Propulsion Laboratory who has done navigation work for several Mars missions, including the Mars Science Laboratory, the ambitious mission that brought the rover *Curiosity* to the red planet in 2012. Nor is there any GPS for the solar system, he adds, which would be a useful way to correct for the effects of solar radiation—just one of the many things that can send a spacecraft off-course. The only GPS we have is on Earth, so we've harnessed it for space travel. With no objective positioning system anywhere else out there, Earth becomes the reference point for a journey to Mars. "The symbol I like to use is that we navigate the spacecraft by looking in the rearview mirror," Martin-Mur explains.

Of all the spacecraft bound for Mars, fewer than half have completed the journey, so great is the number of things that can go wrong. The "entry descent landing," from the first moment of entry into the Martian atmosphere until the moment of touchdown on the surface, seven minutes later, involves a needle-threading of almost unfathomable complexity. The beginning of the descent marks the complete disconnection of the mission from earthly control. From that point on, everything is automated.

The process of getting the 2,000-pound *Curiosity* rover to the ground was like something that sprang from the mind of Rube Goldberg: the spacecraft began its descent at a speed of 13,000 miles per hour, its heat shield reaching a temperature of 1,600 degrees Fahrenheit as it slowed to 1,000 mph, at which time the heat shield ejected, at an angle that ensured it would miss the parachute that opened simultaneously; at 200 mph, the chute came off, as rockets propelled the vehicle to the side to avoid hitting it; 20 meters above the surface, *Curiosity* descended on a tether, and its mothership flew off as soon as the rover landed, to avoid a collision.

None of that had any chance of working if *Curiosity* did not enter the Martian atmosphere at exactly the prescribed point, at exactly the planned angle, at exactly the right speed, at exactly the right time. The Mars Climate Orbiter, launched in 1998, had failed because mission controllers were confused about whether some data was in metric or nonmetric units, causing the thrusters to fire incorrectly. Although the errors were small, they threw the trajectory off by hundreds of kilometers, causing the Orbiter to obliterate itself when it entered the atmosphere. The following year, NASA launched the Mars Polar Lander, which failed in two ways: it didn't reach the surface of Mars, and the two probes it ejected over the planet's south polar region were never heard from again. Two years passed before the next mission, Mars Odyssey, which was sent to orbit the planet, not land on it. A slightly less daunting task, maybe, but NASA wasn't taking any chances—or rather, it found a way to narrow the degree of chance.

To mark the position, speed, and trajectory of spacecraft headed for Mars, NASA has, for many years, relied on its Deep Space Network (DSN): three facilities—in Spain, Australia, and California's Mojave Desert—that house giant dish-shaped antennas, some with a diameter that would almost accommodate a twenty-story building. The oldest technique is to use DSN to measure Doppler shift, the method adopted by the Transit satellites. By 1977, when NASA launched the Voyager probes, the first spacecraft to explore past Mars, this watershed journey demanded additional precision. The long trip would magnify any navigational errors, and the course heading when they reached Saturn—roughly on the same plane as the Earth's equator—made Doppler data much less effective. One solution was two-way ranging, bouncing a signal off the spacecraft and back and using the elapsed time to gauge distance.

That still wasn't enough precision for Voyager, so NASA added a third method: Delta-Differential One-Way Ranging, a.k.a. Delta-DOR or ΔDOR, a modified version of VLBI. The staff at two DSN locations measure the time it takes the same signal sent from the spacecraft to reach each facility. Calculations based on the arrival times reveal the spacecraft's location, but the resistance of the earth's atmosphere can slow those times, throwing off distance calculations by hundreds of miles. To correct for that error, NASA selects a quasar with a heading close to that of the spacecraft, so that the quasar's signal is encountering the same atmospheric resistance. Since the location of the quasar is already known, NASA can measure the timing error of its signal and apply that same error correction to the spacecraft's signal.

Delta-DOR worked reasonably well for Voyager and some subsequent missions, but throughout the 1980s and 1990s it never reached the 99 percent success rate NASA expects from any Deep Space Network measurements. By the late 1990s, budget cuts had shrunk the Delta-DOR project. Following the Climate Orbiter fiasco in 1998, NASA vowed to enact major changes to the Mars program, including the use of an improved version of Delta-DOR. The Mars Odyssey mission in 2001 marked the first time in twenty years that Delta-DOR was used on a Mars mission. During the six months it took *Odyssey* to reach Mars orbit, Delta-DOR data was used to make necessary course corrections, and the technique showed a marked improvement.

The secret weapon was GPS. Quasars may be stable, from our perspective, but our planet is not. The earth's wobble and its polar motion—the shifting of the orientation of its rotational axis, which causes the North Pole to wander about 12 miles every year—skew distance calculations. These effects were somewhat predictable and could be modeled into calculations, but there was still noisy volatility. A low-pressure system whipping across the Great Plains might hurl itself into the Rocky Mountains, causing planetary torque and affecting the earth's rotation speed. Ocean tides and currents are constantly causing subtle planetary deformations. Our ability to harness GPS had grown powerful enough to quiet some of this noise.

Next to every antenna at a Deep Space Network site stands a little black box that cradles a dual-frequency GPS receiver. It receives the same civilian GPS signal as every smartphone in the world. As precise as VLBI is, any facility that uses it is still rooted to the earth, subject to all the planet's movements and deformations. The immensely powerful radio telescopes, capable of measuring radio transmissions from a stable quasar, are themselves slipping and sliding, in ways almost imperceptible but enough to affect measurements. Those GPS receivers are referenced to the earth's center—which, though unstable, is under constant surveillance by the global GPS infrastructure, from the Jet Propulsion Laboratory, Schriever Air Force Base, the National Geospatial-Intelligence Agency, the global web of satellite monitoring stations, and the government agencies, universities, and research institutions worldwide that contribute the data that maintains the International Terrestrial Reference Frame.

The monstrous Deep Space Network antennas shift position in response to small variations in the earth's rotation rate and rotational axis. GPS data from a vast global network of ground terminals enables a continuous real-time assessment of these movements. This information is crucial for strengthening Delta-DOR calculations, which in turn improve deep space navigation. GPS pinpoints the location and orientation of those antennas in what physicists call inertial space, a stable frame and set of coordinates used to define a spacecraft's movement. "It used to be when you were three days out from Mars, you would have to do your last correction, because it would take several days to process the data," says Stephen Lichten, manager of the Jet Propulsion Laboratory's Communications, Tracking, and Radar Division. "But now, with GPS, you can get the data instantly, and you can make corrections much closer to the actual arrival."

Like any spacecraft that enters the Martian atmosphere, the Mars Science Laboratory required seven minutes to land on the surface. Because the signals from a spacecraft require fourteen minutes to make the long journey back to Earth, by the time NASA learned that *Curiosity* had reached the atmosphere, the point beyond which it cannot be controlled, its fate had been sealed for seven minutes, and it would be seven more minutes before NASA knew if the mission had succeeded. NASA engineers call this period "seven minutes of terror," when all they can do is wait.

History shows that the wait ends in dismay more often than elation. But *Curiosity* hit its mark on Mars nearly perfectly, beginning the world's most ambitious exploration of the planet. "This could not have been done thirty years ago," Martin-Mur says. "Because we didn't have navigation that was accurate, and we didn't have all these corrections. There was a time when having errors in meters was OK. Now we have errors of centimeters or millimeters."

Someday, maybe, we won't have to navigate away from Earth by looking through the rearview mirror. Maybe some Galactic Positioning System will let us look through the windshield with confidence. For now, we make do with an elaborate home-centered system, like the first humans who, knowing nothing about navigating by the sky, kept track of their steps in relation to their settlement. We haven't quite reached the *etak* stage yet. Earth is our only reference island. But GPS is our rock.

EPILOGUE

Direction Home

We now know what Tupaia was trying to tell us.

In 2007, two scholars from the Center for Research and Documentation on Oceania in Marseille, France, published an analysis of the document now held at the British Museum, labeled "Chart of the Society Islands with Otaheite [Tahiti] in the center July–Aug 1769"—otherwise known as Tupaia's map. They began by choosing a few islands that Tupaia would have considered central to his life, islands from which he was likely to begin a journey. They used a modern nautical chart to compute the bearings from those islands to several destination islands, superimposed those sets of bearings on Tupaia's document, and rotated them to see if they correlated with the bearings depicted on the map. In most cases, they did.

The results suggested that Tupaia had not created an objective map of the Pacific, but a subjective view from his perspective presented, as closely as he could imagine, in the language of Cartesian space. The map was not so much a map as it was a "mosaic of sailing directions." It was a way for Tupaia to represent his conception of navigating the Pacific. Tupaia was trying to map a system similar to *etak*. He was attempting to reconcile two very different worldviews, and he nearly succeeded. His map is relatively accurate *as a map*—enough to make Cook think that's what it was, but not enough for it to make sense to him. "Both could look at the manuscript and see their own system represented," the researchers noted.

By the time this analysis appeared, the mystery of Polynesian migration was largely settled. Captain Cook's initial inclination to believe in an easterly migration was correct. The most powerful circumstantial evidence began as a project envisioned by Ben Finney, an Australian National University colleague of David Lewis, the man who had studied navigation under Tevake. A young anthropologist whose deep tan and tousled hair made him look like a slacker surfer, Finney was obsessed with the idea of undertaking a 2,000-mile voyage between Hawai'i and Tahiti, using only the materials and methods of the original voyagers. Tahiti was considered an ancestral homeland, and there was evidence of voyages to and from both places, but these contacts had been severed for at least 500 years, after which the Hawaiians became mostly coastal voyagers. If a sailing canoe could survive that journey, it would go a long way toward proving not only that Polynesians were capable of long-distance navigated voyages, but that they were able to make the journey in both directions.

Finney kept the idea a secret for several years, worried that the pipe-dream project would damage his nascent academic career. The voyage took several years to plan. Finney and his like-minded researchers wanted to recreate with maximum accuracy the look and material of an ancient sailing canoe. After extensive research, they built a 60-foot vessel from plywood and laminated oak.

The "rules" of the project stipulated that no modern technology could be used for navigation, to reproduce as closely as possible the conditions faced by the original voyagers. Knowledge of traditional navigational methods had long died out among native Hawaiians. Finney recruited Lewis to help him find a navigator from elsewhere in Polynesia, but Lewis felt sure there were none. His friend Tevake was the closest he had ever encountered. But if they were willing to expand their search a bit outside the Polynesian Triangle, Lewis said, west to Micronesia, there were some isolated atolls in the Carolines where some navigators still used *etak*, a system that was probably similar to what the Hawaiians had used.

As it happened, Finney knew someone who had married the niece of one of the region's greatest navigators. His name was Pius Piailug, though everyone called him Mau, and he hailed from an atoll called Satawal. Mau agreed to serve as the voyage's navigator, but cautioned that the trip lay far outside the region of the world he knew best. Most of the atolls in his region were separated by less than 100 miles of ocean. Navigating an inter-atoll voyage would typically involve *etak* segments that were between 10 and 20 miles. In the vast space between Hawai'i and Tahiti, there are very few adjacent islands, so for reference Mau would need to use large distant archipelagos, such as the Marshall Islands, 2,000 miles west of the course line, and the Marquesas, hundreds of miles to the east. The *etak* segments would be very long, which meant hundreds of miles guided by a single horizon star.

The vessel, named *Hokule'a*, launched in 1976. Mau utilized his navigation system. As a backup, Lewis supplemented Mau's calculations with Western forms of dead reckoning, such as mentally timing bubbles that moved past the boat to compute the canoe's speed, and using that data to estimate latitude and longitude. After a tense month at sea, *Hokule'a* reached Papeete, the capital of Tahiti, where it was welcomed by a crowd of thousands. The project was a success, demonstrating that early Polynesians were likely adept enough to navigate over long distances. For Mau, however, the experience was a letdown. Disillusioned by the behavior and attitude of some of the crew during the trip, he refused to make the return trip, so *Hokule'a* sailed to Hawai'i without him.

Two years later, an attempt to repeat the voyage ended in disaster. A few hours after leaving Honolulu, *Hokule'a* capsized, killing one crew member, a well-known local surfer named Eddie

Aikau, who attempted to paddle in search of help and was never seen again. A year later, a repaired *Hokule'a* once again set out for Tahiti. Its head navigator had been a member of the second, ill-fated crew, a young part-Native Hawaiian man named Nainoa Thompson. He had no prior navigational experience and knew he'd never master *etak*, a system that demands many years of preparation and training. For the third voyage, he immersed himself in studying the stars, and developed a modified navigational system that used aspects of *etak* while remaining idiosyncratically his own creation.

Though Mau had sworn off participating in any more of *Hokule'a*'s activities, he was impressed enough by Thompson's devotion to agree to return to Hawai'i to help train him. One day in November 1979, they met for their final session before the voyage, which would see *Hokule'a* successfully reach Tahiti once again. Mau and Thompson drove to a lookout point on the eastern edge of Honolulu. The Pacific stretched out into the seemingly infinite horizon. The Hawaiian islands of Molokai and Lanai were faintly visible, but Mau wanted Thompson to see beyond them, all the way to his destination.

Mau asked Thompson to point to Tahiti. Unlike Tupaia on Cook's ship two centuries earlier, Thompson hesitated. How could he pinpoint, with little forethought, a speck in the ocean 2,000 miles away? He thought carefully. "I cannot see the island," he replied. "But I can see an image of the island in my mind."

Mau nodded. "Good. Don't ever lose that image, or you will be lost."

The exam was over. "Let's get in the car," Mau said. "Let's go home."

Notes

Pinpoint is the product of both primary and secondary research, including unpublished interviews conducted by the author between 2012 and 2015. Material from those interviews is not sourced here, except in instances where clarity requires it.

Introduction: The Whisper from Space

xiii "greatest Air Force Base": Col. John Shaw, "The Greatest Air Force Base the World Has Never Seen," August 30, 2011, http://www.afspc.af.mil/news1/story.asp?id=123269990.

xv A pinpoint calculation: For a good concise explanation, see http://www.trimble.com/gps_tutorial/howgps-triangulating.aspx.

xvi Nearly 3 billion mobile apps . . . more than double: European Global Navigation Satellite Systems Agency, "GNSS Market Report," issue 4, March 2015.

xvi "so large": Len Jacobson, *GNSS Markets and Applications* (Boston: Artech House, 2007), 61–2.

xviii "world's only global utility": http://www.schriever.af.mil/GPS/.

xix account for Einstein's laws of relativity: Sameer Kumar and Kevin B. Moore, "The Evolution of Global Positioning System Technology," *Journal of Science Education and Technology* 11, no. 1 (March 2002): 59–73.

Chapter 1: Tupaia Goes Home

3 Around 250 million years ago: "Continents In Collision: Pangaea Ultima," *NASA Science News*, October 6, 2000, http://science1.nasa.gov/science-news/science-at-nasa/2000/ast06oct_1/.

4 The watery distance: On Austronesian migration, see Ben Finney, "The Other One-Third of the Globe," *Journal of World History* 5, no. 2 (1994): 273–97; Will Kyselka, *An Ocean In Mind* (Honolulu: University of Hawai'i Press, 1987).

5 It is difficult to overstate: Jared Diamond has called the establishment of Aus-

tronesian language and culture in Madagascar "the single most astonishing fact of human geography for the entire world." Quoted in A. Kumar, "The Single Most Astonishing Fact of Human Geography," *Indonesia* 92 (2011): 59–95.

5 last great premodern human migration: O. K. H. Spate, "The Pacific as an Artefact," in *The Changing Pacific: Essays In Honor of H. E. Maude*, edited by Neil Gunson (New York: Oxford University Press, 1978).

5 They crossed the ocean in canoes: Finney, "The Other One-Third of the Globe," 279.

5 European navigators were still wary: Harold Gatty, *Nature Is Your Guide: How to Find Your Way On Land and Sea* (London: Collins, 1958).

5 500,000 souls: Ben Finney, "Myth, Experiment, and the Reinvention of Polynesian Voyaging," *American Anthropologist* 93, no. 2 (1991): 383–404, esp. 395.

6 In the early sixteenth century: Spate, "The Pacific as an Artefact."

6 "The Pacific no longer appeared": Quoted ibid.

7 fewer than 500 European ships: Ibid.

7 "the most extensive Nation": Quoted in Finney, "The Other One-Third of the Globe."

7 "How shall we account": James Cook, *The Journals of Captain Cook: Prepared from the Original Manuscripts by J. C. Beaglehole for the Hakluyt Society, 1955–67* (London: Penguin Books, 2003), ccclxvi.

7 "the Machiavelli": Joan Druett, *Tupaia: Captain Cook's Polynesian Navigator* (Santa Barbara, CA: Praeger, 2011), xii.

8 *arioi*, an elite society: Anne Salmond, *The Trial of the Cannibal Dog: Captain Cook in the South Seas* (New Haven, CT: Yale University Press, 2003), 37.

8 "in search of what chance": Ibid., 107.

9 "These people sail": Ibid., 105.

10 "the above list": Druett, *Tupaia*, 121

10 "Shrewd, Sensible, Ingenious": Cook, *The Journals of Captain Cook*, 189–90.

11 *Now I am still alive*: David Lewis, *We, the Navigators: The Ancient Art of Landfinding in the Pacific* (Honolulu: University of Hawai'i Press, 1994), 55.

11 "the first Polynesian navigator": Ibid., 31.

11 "almost alone": Ibid.

11 "I became his pupil": Ibid.

12 the mysteries of Polynesian migration: On the history of the various Polynesian navigation theories, see Ben Finney, *Voyage of Rediscovery: A Cultural Odyssey Through Polynesia* (Berkeley: University of California Press, 1994), 18–23; and Finney, "Myth, Experiment, and the Reinvention of Polynesian Voyaging."

12 "Most people believe": Andrew Sharp, *Ancient Voyagers in Polynesia* (Berkeley: University of California Press, 1964), 53.

13 "would be nothing for a seaman of his caliber": Lewis, *We, the Navigators*, 355–6.

14 The Polynesian navigator's primary tool: For an excellent concise overview of Polynesian navigation techniques, see Oliver Kuhn, "Polynesian Naviga-

tion," *CSEG Recorder*, September 2008, http://csegrecorder.com/features/view/science-break-200809.

17 "embed the route": Reginald G. Golledge, "Human Wayfinding and Cognitive Maps," in *Wayfinding Behavior: Cognitive Mapping and Other Spatial Processes*, edited by Reginald G. Golledge, 5–45 (Baltimore: Johns Hopkins University Press, 1999), 6–7. On the concept of wayfinding, see also Reginald G. Golledge and Tommy Gärling, "Cognitive Maps and Urban Travel," in *Handbook of Transport Geography and Spatial Systems*, edited by David A. Hensher, Kenneth J. Button, Kingsley E. Haynes, and Peter Stopher, 501–12 (Oxford: Emerald Group, 2004): "Navigation implies that a route to be followed is predetermined, is deliberately calculated . . . and defines a course to be strictly followed between a specified origin and destination. . . . Wayfinding is taken more generally to involve the process of finding a path (not necessarily previously traveled) in an actual environment between an origin and a destination that has previously not necessarily been visited."

17 He contrasted home-center systems: Gatty, *Nature Is Your Guide*, 47. Gatty's writing on wayfinding is discussed in Kuhn, "Polynesian Navigation," and in Lewis, *We, the Navigators*.

19 "into which the navigator's": Quoted in Lewis, *We, the Navigators*, 179.

19 "Everything passes by": Ibid., 176. For Lewis's overview of *etak*, see 173–6.

20 "It is easy for us to forget": Ibid., 184.

22 Many who have pondered the mysteries: A few examples: Cook's biographer, Anne Salmond: "Unfortunately, neither Cook nor Banks recorded any of Tupaia's descriptions of the islands, their sailing directions, or the expeditions which had peopled them" (*The Trial of the Cannibal Dog*, 111). Tupaia's biographer, Joan Druett: "[I]nstead of listening to Tupaia's lyrical descriptions of what awaited in the west, and asking questions about how he would make his course there, he and Banks put Tupaia to work answering questions for the reports they were writing about the manners and customs of the Tahitian people" (*Tupaia*, 117). David Lewis: "[N]o one seems to have asked [Tupaia] how he oriented himself, or what his actual concepts and methods were" (*We, the Navigators*, 9).

Chapter 2: The When and the Where

24 the transit of Venus: Steven Cherry, *Transit of Venus: The Other Half of the Longitude Story*, Techwise Conversations, n.d., http://spectrum.ieee.org/podcast/geek-life/profiles/transit-of-venus-the-other-half-of-the-longitude-story.

26 "discovering the longitude": Dava Sobel, *Longitude: The True Story of a Lone Genius Who Solved the Greatest Scientific Problem of His Time* (New York: Walker and Co., 1995), 56.

27 In 1800, Chevalier de Lamarck: Walter Sullivan, "The IGY—Scientific Alliance In a Divided World," *Bulletin of the Atomic Scientists* 14, no. 2 (February 1958): 68–72.

28 first international organization: Ibid., 68.

28 "the largest organized intellectual enterprise": John A. Simpson, "The International Geophysical Year—A Study of Our Planet," *Bulletin of the Atomic Scientists* 13, no. 10 (December 1957): 351.

29 "Contemplate the satellite," "that we puny people": Daniel Lang, "Earth Satellite No. 1," *New Yorker*, May 11, 1957.

29 "the greatest boon": Ibid.

29 "This is our first step": Ibid.

30 "I wouldn't care": "National Affairs: PROJECT VANGUARD," *Time*, October 21, 1957.

30 "I wish to congratulate," "really fantastic": C. M. L. Green and M. Lomask, *Vanguard: A History*, NASA SP-4202, NASA Historical Series (Washington, DC: National Aeronautics and Space Administration, 1970), http://www.spacearium.com/filemgmt_data/files/Vanguard_A_History.pdf, linked comment.

30 a golf ball ejected by a jet plane: Green and Lomask, *Vanguard*.

32 The countdown began: For details of TV-3's launch, see Milton Bracker, "Vanguard Rocket Burns on Beach; Failure to Launch Test Satellite Seen as Blow to U.S. Prestige," *New York Times*, December 7, 1957.

34 NYSE officials temporarily suspended trading: "Martin's Stock Falls As Its Vanguard Fails," *New York Times*, December 7, 1957.

34 "It wasn't a long flight": Bracker, "Vanguard Rocket Burns."

34 the nation's editorial writers: all editorial excerpts are from "Editorial Comment on the Nation's Failure to Launch a Test Satellite," *New York Times*, December 8, 1957.

35 "full, complete": Green and Lomask, *Vanguard*.

35 a Soviet program: "Russians at U.N. Tweak U.S. on (Satellite) Nose," *New York Times*, December 7, 1957.

38 "it occurred to me": memo reproduced in R. J. Danchik, "An Overview of Transit Development," *Johns Hopkins APL Technical Digest* 19, no. 1 (1998): 19, http://techdigest.jhuapl.edu/TD/td1901/danchik.pdf.

38 within a half mile: memo reproduced ibid.

38 The Transit system was fully operational: For details on Transit's use, see ibid.; R. B. Kershner and F. T. McClure, "The Legacy of Transit: A Dedication," *Johns Hopkins APL Technical Digest* 19, no. 1 (1998): 3, http://techdigest.jhuapl.edu/td/td1901/pisacane.pdf.

38 "the largest step": Danchik, "An Overview of Transit Development."

39 a second radio fence: For details, see Roger L. Easton, "Keynote Address: In the Beginning of GPS," *Proceedings of the Thirty-Second Annual Precise Time and Time Interval (PTTI) Meeting*, Reston, VA, 2000.

40 "not general enough . . . feasible": Roger L. Easton, "A Satellite Navigation Sys-

tem," Naval Research Laboratory, Space Surveillance Branch Technical Memorandum no. 112, 1964. Special thanks to Richard Easton for providing this document.

40 You now have tools: For Easton's description of passive ranging, see L. B. Slater, "From Minitrack to NAVSTAR: The Early Development of the Global Positioning System, 1955–1975," paper delivered at 2011 IEEE/MTT-S International Microwave Symposium, 2011.

41 within the clock itself: Tony Jones, *Splitting the Second: The Story of Atomic Time* (Bristol, U.K., and Philadelphia: Institute of Physics, 2000), 33.

41 settled on caesium: Ibid., 46.

41 quartz oscillator: Ibid., 83–4.

42 "Some time was spent . . . ones": Roger L. Easton, "An Exploratory Development Program In Passive Navigation," Naval Research Laboratory, Space Applications Branch Technical Memorandum no. 1, May 8, 1967.

43 Timation I: R. L. Beard, J. Murray, and J. D. White, "GPS Clock Technology and the Navy PTTI Programs at the U.S. Naval Research Laboratory," 1986, http://www.dtic.mil/cgi-bin/GetTRDoc?AD=ADA492721&Location=U2&doc=GetTRDoc.pdf.

44 a few ten-millionths of a second: For details on the time-transfer experiment, see Jay Oaks, and James A. Buisson, "Satellite Time Transfer Past and Present," *Proceedings of the Thirty-Fourth Annual Precise Time and Time Interval (PTTI) Systems and Applications Meeting,* Reston, VA, 2002, NASA Conference Publications 3159, 7–18.

44 Timation II: Brad Parkinson and Stephen T. Powers, "The Origins of GPS," *GPS World*, June 2010, 38.

44 Responding to a directive: Ron Beard, author interview; and Ron Beard, "Another View of GPS Origins," *GPS World*, December 2010.

45 "literally hundreds": Parkinson and Powers, "The Origins of GPS."

Chapter 3: Global Reach, Global Power

49 doctrine of high-altitude precision bombing: On the evolution of thinking regarding precision bombing in the Air Force, see Michael Russell Rip and James M. Hasik, *The Precision Revolution: GPS and the Future of Aerial Warfare* (Annapolis, MD: Naval Institute Press, 2002), 34–5, 45–6.

51 computerized firing system: On the gunship concept and the development of the AC-130, see Jack S. Ballard, *Development and Employment of Fixed-Wing Gunships, 1962–1972* (Washington, DC: Office of Air Force History, United States Air Force, 1982), 84–5.

52 "the automated battlefield": Quoted in Frank Barnaby, Ronald Huisken, and Stockholm International Peace Research Institute, *Arms Uncontrolled* (Cambridge, MA: Harvard University Press, 1975), 72.

52 Igloo White: Phil Stanford, "The Automated Battlefield," *New York Times*, February 23, 1975.

52 "keen appreciation": Brad Parkinson, author interview.

54 "I recognized that there was pressure": Brad Parkinson, oral history interview by Michael Geselowitz.

54 On Labor Day 1973: On the "lonely halls" meeting, see Bradford W. Parkinson and Stephen T. Powers, "Fighting to Survive: Five Challenges, One Key Technology, the Political Battlefield—and a GPS Mafia," *GPS World*, June 2010.

54 the actress Hedy Lamarr: Len Jacobson, *Flying For GPS* (Xlibris, 2014), 34.

55 This is what every GPS receiver: For a concise explanation of GPS and spread-spectrum signals, see Jacobson, *Flying For GPS*, 46–8.

55 10-watt bulb in Rome: Jules McNeff, "The Global Positioning System: A Quiet Revolution in Time and Space," *IEEE Transactions on Microwave Theory and Techniques* 50, no. 3 (March 2002).

55 amplify the signal: "Using Psuedo [sic] Random Code as an Amplifier," http://www.trimble.com/gps_tutorial/sub_amplify.aspx.

56 "otherwise brilliant career": Brad Parkinson, "GPS For Humanity," Stanford Engineering Hero Lecture, 2012, http://www.youtube.com/watch?v=d6I6wFf-X_c.

56 some of Easton's Timation colleagues: Richard D. Easton and Eric F. Frazier, *GPS Declassified: From Smart Bombs to Smartphones* (Lincoln, NE: Potomac, 2013), 64–8.

57 just to meet with themselves: Ibid., 64; Ron Beard, author interview.

58 a sign on the wall of his office: Parkinson and Powers, "Fighting to Survive," 17.

58 "I had great sensitivity": "Our History—In Their Own Words: Dr. Bradford Parkinson," interview by Steven R. Strom, June 1, 2003, http://www.aerospace.org/about-us/history/in-their-own-words/dr-bradford-parkinson/ (quotation edited slightly for coherence).

60 a doctrine of air superiority: Dana J. Johnson, "Overcoming Challenges to Transformational Space Programs: A Case Study of the Early History of the Global Positioning System (GPS)," in *American National Security Policy: Essays in Honor of William R. Van Cleave*, edited by Bradley A. Thayer (Fairfax, VA: National Institute Press, 2007).

60 The GPS receiver that Collins developed: Ibid.

60 a report that was highly critical: Matthew E. Skeen, "The Global Positioning System: A Case Study in the Challenges of Transformation," *JFQ: Joint Force Quarterly* 51 (Autumn 2008): 88–93.

61 Operation El Dorado Canyon: Rip and Hasik, *The Precision Revolution.*

63 Col. Roland Ellis: Donald J. Kutyna, "Indispensable: Space Systems in the Persian Gulf War," *Air Power History* 46, no. 1 (Spring 1999): 28.

63 The Air Force struggled: Michael P. Scardera, "NAVSTAR GPS Operations," in *From the Line in the Sand: Accounts of USAF Company Grade Officers in Support of Desert Shield/Desert Storm*, edited by Michael P. Vriesenga, 123–35 (Maxwell AFB, AL: Air University Press, 1994).

63 Chuck Horner was fond: Tom Clancy and Chuck Horner, *Every Man a Tiger: The Gulf War Air Campaign* (New York: Putnam, 1999), 34–5.

63 He had visited: Ibid., 178–80.

64 "you cannot trust": Ibid., 99.

64 "stupid, evil, aimless": Ibid., 96–8.

64 Glosson also took inspiration: Buster Glosson, *War with Iraq: Critical Lessons* (Charlotte, NC: Glosson Family Foundation, 2003), 115.

64 "We never had a clue": Ibid., 109.

65 "It doesn't make any difference": Gaylord Green, author interview.

67 A squadron of thirteen B-52s: Rip and Hasik, *The Precision Revolution*, 152.

67 A U-2 reconnaissance: Rip and Hasik, *The Precision Revolution*, 145–6.

67 the first large-scale deep desert advance: James M. Hasik and Michael Russell Rip, "GPS at War: A Ten-Year Retrospective," *Proceedings of the Fourteenth Annual International Technical Meeting of the Satellite Division of the Institute of Navigation*, Salt Lake City, UT, 2001 (ION GPS 2001), 2406–17.

67 unexploded mines: Kutyna, "Indispensable," 28.

67 total number of GPS units: Rip and Hasik, *The Precision Revolution*, 135–6, 187.

68 "came of age": Thomas S. Moorman, Jr., "Space: A New Strategic Frontier," *Airpower Journal* 6, no. 1 (Spring 1992), http://www.airpower.maxwell.af.mil/airchronicles/apj/apj92/spr92/moor.htm.

68 Russia's most prominent military scientist: Rip and Hasik, *The Precision Revolution*, 131.

68 At 3 a.m.: For details of Operation Secret Squirrel, see John Tirpak, "The Secret Squirrels," *Air Force Magazine*, April 1994; also Rip and Hasik, *The Precision Revolution*, 159–61.

69 The U.S. did not confirm: Jon Lake, *B-52 Stratofortress Units in Operation Desert Storm* (Oxford: Osprey, 2004), 31.

71 The JDAM officially entered: For details of the B-2 flight, see Rip and Hasik, *The Precision Revolution*, 245–6.

72 as fighting intensified in Afghanistan: John Hendren and Maura Reynolds, "The US Bomb That Nearly Killed Karzai," *Los Angeles Times*, March 27, 2002, http://articles.latimes.com/2002/mar/27/news/mn-34908.

72 A recent survey: Don Jewell, "A Guide to Trends in GPS/PNT User Equipment," presented at the eleventh meeting of the PNT Advisory Board, May 7, 2013, http://www.gpsworld.com/trends-in-gpspnt-user-equipment/.

Chapter 4: Ranging the Perfect Beet

73 "land of starvation": quoted in Eric Twitty, "Silver Wedge: The Sugar Beet Industry in Fort Collins," SWCA Cultural Resource Report, SWCA Environmental Consultants, August 2003, submitted to Advance Planning Department, City of Fort Collins, CO.

74 A half century after: On the early history of the Front Range beet industry, see

Llewellyn Alexander Moorhouse, *Farm Practice in Growing Sugar Beets for Three Districts in Colorado 1914–15* (Washington, DC: U.S. Department of Agriculture, 1918).

74 Sugar factories had a catalytic effect: On the economic effects of the sugar industry in the region, see Dena S. Markoff, "The Beet Sugar Industry in Microcosm: The National Sugar Manufacturing Company, 1899 to 1967," *Journal of Economic History* 41, no. 1 (1981): 193–5.

74 "the cultivation of beets": "Fort Collins History Connection: Fort Collins History and Architecture," http://history.fcgov.com/archive/contexts/sugar.php.

74 Adjusted for inflation . . . improved agricultural practices: William Jon May, *The Great Western Sugarlands: The History of the Great Western Sugar Company and the Economic Development of the Great Plains* (New York: Garland, 1989).

76 The Air Force polled 500: Jacobson, *GNSS Markets and Applications*, 64.

77 The Z-set: John F. Canniff and Christopher B. Duncombe, "Static Evaluation of a NAVSTAR GPS (Magnavox Z-Set) Receiver (May–September 1979): Final Report," (Washington, DC: U.S. Department of Transportation, Research and Special Programs Administration, May 1980).

77 Operation Eagle Claw: Jacobson, *Flying For GPS*, 54–5.

77 Hewlett-Packard history: Michael S. Malone, *Bill & Dave: How Hewlett and Packard Built the World's Greatest Company* (New York: Portfolio, 2007).

78 the "HP way": Ibid.

78 the size of its workforce doubled: Ibid

82 The plane had drifted: Eliot Marshall, "Steering Clear of Sakhalin," *Science* 222, no. 4621 (October 21, 1983): 303–4.

82 "World opinion is united": Reagan's executive order reproduced in Rip and Hasik, *The Precision Revolution*, 429–30.

82 calls in Congress: Marshall, "Steering Clear of Sakhalin," 303–4.

83 "official U.S. user" status: Michael J. Ellett, "Civil Access to the Precise Positioning Service of the NAVSTAR Global Positioning System," *Proceedings: IEEE Position Location and Navigation Symposium '86* (Institute of Electrical and Electronics Engineers, 1986).

86 "Navigation used to be deadly": William F. Buckley, "Precision Sailing," *New York Times Magazine*, May 19, 1985.

88 Ashtech made history: Javad Ashjaee, "How GPS and GLONASS Got Together—and Other Recent Events," *GPS World*, June 2011, 60–6.

96 producing around 4,000: On the number of SLGRs and Magellans in Desert Storm, and their specifications, see Breck W. Henderson, "Ground Forces Rely on GPS to Navigate Desert Terrain," *Aviation Week & Space Technology*, February 11, 1991. On Trimble's production problems: Charlie Trimble, author interview.

96 During the run-up to Desert Storm: Scott Pace, author interview.

97 "far more accurate": Quoted in "Pentagon's Rigid Position on Positioning," *Bulletin of the Atomic Scientists* 47, no. 7 (September 1991), 4.

97 "The Pentagon apparently": Ibid.

100 "any enemy of the United States": Aeronautics and Space Engineering Board, National Research Council, *The Global Positioning System: A Shared National Asset* (Washington, DC: National Academies Press, 1995).

100 worth $2 billion in 1996: Jacobson, *GNSS Markets and Applications*, 97.

100 "This will mean": "Statement by the President Regarding the United States' Decision to Stop Degrading Global Positioning System Accuracy," May 1, 2000, http://clinton3.nara.gov/WH/EOP/OSTP/html/0053_2.html.

100 Magellan's retailers: Amy Gilroy, "Enhanced GPS Accuracy Boosts Consumer Interest," *TWICE: This Week in Consumer Electronics* 15, no. 14 (June 12, 2000): 34.

101 "supporting GLONASS": Ashjaee, "How GPS and GLONASS Got Together."

104 Between 2006 and 2012: European Global Navigation Satellite Systems Agency, "GNSS Market Report."

104 $33 billion: Nam D. Pham, "The Economic Benefits of Commercial GPS Use in the US and the Costs of Potential Disruption," NDP Consulting, June 2011, http://saveourgps.org/pdf/GPS-Report-June-22-2011.pdf. The study analyzed U.S. crop production from 2007 to 2010 and estimated that GPS reduced input costs by $9.8 billion and accounted for $10.1 billion in output.

104 "there is a long way to go": "Precision Agriculture: An Opportunity for EU Farmers," European Parliament, Directorate-General For Internal Policies, Policy Department B, Structural And Cohesion Policies, Agriculture and Rural Development, 2014, 35–6.

104 By 2020: "Precision Farming Market by Technology (GPS/GNSS, GIS, Remote Sensing & VRT), Components (Automation & Control, Sensors, FMS), Application (Yield Monitoring, VRA, Mapping, Soil Monitoring & Scouting) and Geography—Global Forecasts to 2020," MarketsandMarkets, 2014.

104 nearly 50 percent of the world's tractors: European Global Navigation Satellite Systems Agency, "GNSS Market Report."

104 The fastest-growing market: European Global Navigation Satellite Systems Agency, "GNSS Market Report."

105 Evidence suggests: Caspar Van Vark, "From Agribusiness to Subsistence: High-Tech Tools Now Available to All," *Guardian*, June 4, 2014.

105 In Uttar Pradesh: Ibid. See also Madeline Fisher, "Precision Ag in the Developing World," *CSA News*, February 2012.

106 "This circumstance very well": Quoted in Finney, *Voyage of Rediscovery*, 24.

107 chronometer stopped ticking: Sobel, *Longitude*, 151.

Chapter 5: "Death by GPS"

111 a scenic road less traveled: Richard Cockle, "Search Alert For Missing British Columbia Couple Expands to Surrounding States," *Oregon Live*, April 7, 2011.

112 "a pretty good road": Richard Cockle, "Missing Canadian Couple: Experts Say Always Pair GPS with Paper Map, Realize Technology Not Infallible," *Oregon Live*, May 14, 2011.

113 "She came running": Tom Knudson, " 'Death by GPS' in Desert," *Sacramento Bee*, January 30, 2011, http://www.sacbee.com/2011/01/30/3362727/death-by-gps-in-desert.html.

113 the Japanese tourists: Akiko Fujita, "GPS Tracking Disaster: Japanese Tourists Drive Straight into the Pacific," ABC News blogs, April 15, 2013, http://abcnews.go.com/blogs/headlines/2012/03/gps-tracking-disaster-japanese-tourists-drive-straight-into-the-pacific/.

113 in Yorkshire: "Sorry, the Sat-Nav Told Me to Drive Up Here: BMW Left Teetering on 100ft Cliff Edge," *Daily Mail*, March 25, 2009.

113 the woman in Bellevue: "Woman Drives into Swamp Following GPS Directions," *Huffington Post*, June 15, 2011.

113 the Swedish couple: Marc Dorian and Megan Reilly, "My GPS Almost Killed Me," *ABC News*, March 15, 2013, http://abcnews.go.com/Technology/gps-killed-death-valley-debacle-mishaps/story?id=18718677.

114 the elderly woman: Chris Matyszczyk, "GPS Sends Belgian Woman to Croatia, 810 Miles Out of Her Way," *CNET*, January 14, 2013, http://news.cnet.com/8301-17852_3-57563958-71/gps-sends-belgian-woman-to-croatia-810-miles-out-of-her-way/. 1/14./2013.

114 Inuit wayfinding techniques: Claudio Aporta and Eric Higgs, "Satellite Culture: Global Positioning Systems, Inuit Wayfinding, and the Need for a New Account of Technology," *Current Anthropology* 46, no. 5 (December 2005): 729–53. See also Claudio Aporta, "New Ways of Mapping: Using GPS Mapping Software to Plot Place Names and Trails in Igloolik," *Arctic* 56, no. 4 (December 2003): 321–7.

115 Edward Tolman: E. C. Tolman, "Cognitive Maps in Rats and Men," *Psychological Review* 55, no. 4 (1948): 1–11.

116 "making *behavior*": quoted in David E. Leary, "On the Conceptual and Linguistic Activity of Psychologists: The Study of Behavior from the 1890s to the 1990s and Beyond," *Behavior and Philosophy* 32, no. 1 (January 2004): 13–35.

116 such as William James: Tolman's critique of the "telephone switchboard" theorists reads like a rebuke to James's belief that "all our consciousness accompanies a chain of events of which the first was an incoming current in some sensory nerve, and of which the last will be a discharge into some muscle, blood-vessel, or gland"; quoted in David E. Leary, "On the Conceptual and Linguistic Activity of Psychologists: The Study of Behavior from the 1890s to the 1990s and Beyond," *Behavior and Philosophy* 32, no. 1 (January 2004): 17.

117 Tolman argued: On the history of behaviorism in psychology, see Leary, "On the Conceptual and Linguistic Activity of Psychologists."

117 "a map and a picture": Quoted in Laurence D. Smith, "Metaphors of Knowledge and Behavior in the Behaviorist Tradition," in *Metaphors in the History of Psychology*, edited by David E. Leary, 239–66 (New York: Cambridge University Press, 1990), 247.

118 "a map-like representation": Robert M. Kitchin, "Cognitive Maps: What Are

They and Why Study Them?" *Journal of Environmental Psychology* 14, no. 1 (March 1994): 1–19.

118 recent developments in cognitive science: Alex Hutchinson, "Global Impositioning Systems," *The Walrus*, November 2009, http://thewalrus.ca/global-impositioning-systems/.

120 "south-pointing carriage": On the earliest vehicle navigation systems, see Robert L. French, "Automobile Navigation In the Past, Present, and Future," http://mapcontext.com/autocarto/proceedings/auto-carto-8/pdf/pages553-562.pdf; Robert L. French, "Historical Overview of Automobile Navigation Technology," *Proceedings of the Thirty-Sixth IEEE Vehicular Technology Conference* (Institute of Electrical and Electronics Engineers, 1986), 350–8.

121 Nolan Bushnell: Alexis C. Madrigal, "Chuck E. Cheese's, Silicon Valley Startup: The Origins of the Best Pizza Chain Ever," *Atlantic*, July 17, 2013, http://www.theatlantic.com/technology/archive/2013/07/chuck-e-cheeses-silicon-valley-startup-the-origins-of-the-best-pizza-chain-ever/277869/; Ian Bogost, "Persuasion and Gamespace," in *Space Time Play: Computer Games, Architecture and Urbanism: The Next Level*, edited by Friedrich von Boories, Steffen P. Walz, and Matthias Böttger, 304–11 (Basel, Switzerland: Birkhäuser, 2007).

124 If a team were building a robot: Benjamin J. Kuipers, "The Cognitive Map: Could It Have Been Any Other Way?" In *Spatial Orientation: Theory, Research, and Application*, edited by H. L. Pick, Jr., and L. P. Acredolo, 345–59 (New York: Plenum, 1983); Benjamin J. Kuipers and Tod S. Levitt, "Navigation and Mapping in Large-Scale Space," *AI Magazine*, Summer 1988.

124 ideal driver navigational system: David M. Mark and Matthew McGranaghan, "Effective Provision of Navigation Assistance to Drivers: A Cognitive Science Approach," *Proceedings of the Auto-Carto London* (London, 1986), 399–408.

125 well-known psychology experiment: Lynn A. Streeter, Diane Vitello, and Susan A. Wonsiewicz, "How to Tell People Where to Go: Comparing Navigational Aids," *International Journal of Man-Machine Studies* 22, no. 5 (May 1985): 549–62.

125 Another geographer: Scott M. Freundschuh, "Does 'Anybody' Want (Or Need) Vehicle Navigation Aids?", in *Issues in Vehicle Navigation and Information Systems*, edited by Scott M. Freundschuh, Michael D. Gould, and David M. Mark, Technical Report 89-15, National Center For Geographic Information and Analysis, 1989.

126 "What Garmin has done": Justin Ewers, "The Road to Riches," *U.S. News & World Report* 142, no. 3 (January 22, 2007): 59–62.

127 A 2006 German study: Stefan Münzer, Hubert D. Zimmer, Maximilian Schwalm, Jörg Baus, and Ilhan Aslan, "Computer-Assisted Navigation and the Acquisition of Route and Survey Knowledge," *Journal of Environmental Psychology* 26, no. 4 (December 2006): 300–8.

129 "was less effective": Toru Ishikawa, Hiromichi Fujiwara, Osamu Imai, and Atsuyuki Okabe, "Wayfinding with a GPS-based Mobile Navigation System:

A Comparison with Maps and Direct Experience," *Journal of Environmental Psychology* 28, no. 1 (March 2008): 74–82.

129 "attend to objects": Gilly Leshed, Theresa Velden, Oya Rieger, Blazej Kot, and Phoebe Sengers, "In-Car GPS Navigation: Engagement with and Disengagement from the Environment," *Proceedings of the Twenty-Sixth Annual SIGCHI Conference on Human Factors in Computing Systems* (Special Interest Group on Computer–Human Interaction, 2008), 1675–84.

130 "It is as if grid cells: "Human Brain Uses Grid to Represent Space," University College London press release, January 20, 2010, http://www.ucl.ac.uk/news/news-articles/1001/10012001.

130 Other research at UCL: "Scans Reveal How the Brain's GPS Helps Us Navigate from A to B," *Guardian*, November 15, 2011, http://www.guardian.co.uk/science/2011/nov/15/scans-reveal-brain-gps-navigate.

130 "Physical maps help us build": Julia Frankenstein, "Is GPS All in Our Heads?" *New York Times*, February 2, 2012.

130 Wearing head-mounted displays: Julia Frankenstein, Betty J. Mohler, Heinrich H. Bulthoff, and Tobias Meilinger, "Is the Map in Our Head Oriented North?" *Psychological Science* 23, no. 2 (December 29, 2011): 120–5.

132 "The surprising result": Timothy P. McNamara and Christine M. Valiquette, "Remembering Where Things Are," in *Human Spatial Memory: Remembering Where*, edited by Gary L. Allen, 3–24 (Mahwah, NJ: Lawrence Erlbaum Associates, 2004).

132 "The problem with GPS systems": Julia Frankenstein, email to author, March 8, 2012.

132 "The more we rely": Frankenstein, "Is GPS All in Our Heads?"

132 A British study: E. A. Maguire, K. Woollett, and H. J. Spiers, "London Taxi Drivers and Bus Drivers: A Structural MRI and Neuropsychological Analysis," *Hippocampus* 16, no. 12 (2006): 1091–1101, http://onlinelibrary.wiley.com/doi/10.1002/hipo.20233/abstract.

133 mice were trained to run a maze: Hutchinson, "Global Impositioning Systems."

134 "was apparently paying more attention": "Police: GPS May Have Told Couple to Drive Off Cline Avenue Bridge," nwi.com, March 28, 2015, http://www.nwitimes.com/news/local/lake/east-chicago/police-gps-may-have-told-couple-to-drive-off-cline/article_cbd32cd1-bb19-5adb-83a9-ee1e09bd4592.html.

137 In his final hours: On the rescue of Rita Chretien and the discovery of Albert's body: Rita Chretien, author interview; "Remains Of B.C. Man Found In Nevada," *Huffington Post*, January 10, 2012, http://www.huffingtonpost.ca/2012/10/01/albert-chretien-found-nevada-bc_n_1929793.html; Douglas Quan, "'She's an Amazing Woman': Trio Who Found Rita Chretien in the Wilderness Tell Incredible Rescue Story," *Postmedia News*, May 10, 2011; "Missing Canadian Businessman Albert Chretien Found 7 Miles from Stranded Van on Northern Nevada Mountain," *The Oregonian*, October 1, 2012; "Handwrit-

ten Notes Detail Rescued Canadian Woman's Ordeal," CNN, http://www.cnn.com/2011/US/05/09/nevada.missing.couple/index.html; "Missing Canadian Couple," *The Oregonian*, May 14, 2011.

Chapter 6: The Hornet's Nest

139 The last thing a pilot wanted to see: For a cockpit view of the final approach to Juneau International Airport, see http://www.faa.gov/nextgen/snapshots/stories/?slide=10.

139 "Particularly at night": Tomas Kellner, "No Room for Error: Pilot and Innovator Steve Fulton Talks about the 'Alarm and Frustration' that Gave Birth to a Revolution in Aircraft Navigation," *GE Reports*, October 29, 2013, http://www.gereports.com/post/75375269775/no-room-for-error-pilot-and-innovator-steve-fulton.

139 Receiver Autonomous Integrity Monitoring: See Y. C. Lee, "Receiver Autonomous Integrity Monitoring (RAIM) Capability for Sole-Means GPS Navigation in the Oceanic Phase of Flight," *Proceedings: IEEE Position Location and Navigation Symposium '92* (Institute of Electrical and Electronics Engineers, 1992), 464–72.

140 Alaska Airlines introduced RAIM: Per Enge, Nick Enge, Todd Walter, and Leo Eldredge, "Aviation Benefits from Satellite Navigation," *New Space*, September 22, 2014.

140 controlled by the Pentagon: Phillip J. Klass, "Pentagon Urged to Confirm Policy Allowing Civilian Use of GPS Navsats," *Aviation Week & Space Technology*, January 8, 1990.

143 1,600 gallons of fuel: Parkinson, "GPS for Humanity."

143 The FAA estimates: On efficiency improvements for aviation due to GPS, see "NextGen Update: 2014," Federal Aviation Administration, August 2014.

143 "the backbone of our nation's economy": "What Is Critical Infrastructure?" U.S. Department of Homeland Security, http://www.dhs.gov/what-critical-infrastructure. Accessed November 2, 2015.

144 "may be used to augment": "National Space Policy of the United States of America," June 28, 2010, https://www.whitehouse.gov/sites/default/files/national_space_policy_6-28-10.pdf.

144 President Clinton convened: "Critical Foundations: Protecting America's Infrastructures," President's Commission on Critical Infrastructure Protection, October 1997.

145 "permeating the infrastructure": Jacobson 2007, 50.

145 On the railways: European Global Navigation Satellite Systems Agency, "GNSS Market Report."

147 test the porosity of GPS: for a detailed description of VAT's spoofing experiments, see Jon S. Warner and Roger G. Johnston, "A Simple Demonstration

that the Global Positioning System (GPS) Is Vulnerable to Spoofing," *Journal of Security Administration* 25 (2002): 19–28.

148 "Current GPS receivers": Roger G. Johnston and Jon S. Warner, "Think GPS Cargo Tracking = High Security? Think Again," Technical Report, Los Alamos National Laboratory, 2003.

149 "tempting target": John A. Volpe National Transportation Systems Center, "Vulnerability Assessment of the Transportation Infrastructure Relying on the Global Positioning System," Office of the Assistant Secretary for Transportation Policy, U.S. Department of Transportation, August 29, 2001.

151 Humphreys's drone trial: Daniel P. Shepard, Jahshan A. Bhatti, Todd E. Humphreys, and Aaron A. Fansler, "Evaluation of Smart Grid and Civilian UAV Vulnerability to GPS Spoofing Attacks," *Proceedings of the ION GNSS Meeting*, Nashville, TN, 2012; Kyle Wesson and Todd E. Humphreys, "Hacking Drones," *Scientific American* 309, no. 5 (November 2013): 55–9.

153 "well within the capability": Todd E. Humphreys, "Statement on the Vulnerability of Civil Unmanned Aerial Vehicles and Other Systems to Civil GPS Spoofing," submitted to the Subcommittee on Oversight, Investigations, and Management, House Committee on Homeland Security, 2012.

153 the code works: Mark L. Psiaki and Todd E. Humphreys, "GNSS Spoofing and Detection," unpublished paper, 2015.

155 scientists use it to track neutrinos: Stephen Mitchell, "MINOS Timing and GPS Precise Point Positioning," paper presented at the International Workshop on Accelerator Alignment 2012, Fermilab, Batavia, IL.

156 In 1968, the Empress: Nigel Linge, "The Archaeology of Communications' Digital Age," *Industrial Archaeology Review* 35, no. 1 (May 2013): 45–64; UK Telephone History, http://www.britishtelephones.com/histuk.htm. Accessed November 2, 2015.

156 the mostly analog Bell System: On AT&T, time synchronization, and GPS, see Ed Butterline and Sally L. Frodge, "Telecommunications Synchronization and GPS," *GPS Solutions* 5, no. 1 (2001): 51–4. On LAO, see E. W. Butterline, J. E. Abate, and G. P. Zampetti, "Use of GPS to Synchronize the AT&T National Telecommunications Network," Defense Technical Information Center, December 1988.

157 New digital protocols: Phil Mann and Ed Butterline, "Global Positioning System Use in Telecommunications," *Proceedings of the Eleventh International Annual Technical Meeting of the Satellite Division of the Institute of Navigation*, Nashville, TN, 1998 (ION GPS 1998), 1449–54.

159 Italian blackout: Zofia Lukszo, Geert Deconinck, and Margot P. C. Weijnen, *Securing Electricity Supply in the Cyber Age: Exploring the Risks of Information and Communication Technology in Tomorrow's Electricity Infrastructure* (New York: Springer Science & Business Media, 2010), 59–61.

159 better real-time information: A. G. Phadke and J. S. Thorp, "Protection Sys-

tems with Phasor Inputs," *Synchronized Phasor Measurements and Their Applications* (Boston: Springer, 2008), 197–221.

160 the thirty Italy installed: Hassan Bevrani, Masayuki Watanabe, and Yasunori Mitani, *Power System Monitoring and Control* (Hoboken, NJ: John Wiley & Sons, 2014).

160 "revelation": Power Grid Corporation of India, "Unified Real Time Dynamic State Measurement," February 2012, also quoted in John Kemp, "India Experiments with Smart Grid, Too Late," Reuters, August 3, 2012.

161 Humphreys tackled the Synchrophasor problem: Daniel P. Shepard, Todd E. Humphreys, and Aaron A. Fansler, "Evaluation of the Vulnerability of Phasor Measurement Units to GPS Spoofing Attacks," *International Journal of Critical Infrastructure Protection* 5 (December 2012), 146–153; Daniel P. Shepard, Todd E. Humphreys, and Aaron A. Fansler, "Going Up Against Time: The Power Grid's Vulnerability to GPS Spoofing Attacks," *GPS World*, August 2012.

164 "truth reference system": Desiree Craig, "Truth On the Range: USAF's New Reference System," *Inside GNSS*, May/June 2012.

165 "U.S. critical infrastructure": "National Risk Estimate: Risks to United States Critical Infrastructure from Global Positioning System Disruptions," National Protection and Programs Directorate, Department of Homeland Security, November 9, 2011.

165 "ensure that critical infrastructure": U.S. Government Accountability Office, "GPS Disruptions: Efforts to Assess Risks to Critical Infrastructure and Coordinate Agency Actions Should Be Enhanced," November 2013.

166 "the part of the English Channel": Dee Ann Divis, "Proposal for U.S. eLoran Service Gains Ground," *Inside GNSS*, January/February 2014.

167 It might come in handy: European Global Navigation Satellite Systems Agency, "GNSS Market Report."

Chapter 7: Better Living Through Tracking

170 one out of every twenty-five Americans: David Kocieniewski, "Facing the City, Potential Targets Rely on a Patchwork of Security," *New York Times*, May 9, 2005.

171 "the Turnpike is its aorta": Mike Frassinelli, "Exit 13A: A Trip Along the 'Most Dangerous Two Miles in America'," nj.com, October 28, 2011, http://www.nj.com/news/index.ssf/2011/10/exit_13a_how_the_most_dangerou.html.

171 In 2009, Newark Liberty officials: Sam Pullen and Grace Xingxin Gao, "GNSS Jamming In the Name of Privacy," *Inside GNSS*, April 2012, 35–43.

172 when visibility approaches nil: GBAS will likely be certified for category III landings in 2018.

173 "homosexual tendencies": Douglas Robinson, "Delinquents Are Paid by the Hour in Boston to Submit to Study," *New York Times*, December 14, 1964.

174 "a humane technology": Ralph K. Schwitzgebel, *Streetcorner Research: An Exper-*

imental Approach to the Juvenile Delinquent (Cambridge, MA: Harvard University Press, 1964), 103.

174 "behavioral feedback system": Ralph Schwitzgebel, Robert Schwitzgebel, Walter N. Pahnke, and William Sprech Hurd, "A Program of Research in Behavioral Electronics," *Behavioral Science* 9, no. 3 (July 1964): 233–8.

176 If the crime involved: Ralph K. Schwitzgebel and Bernard Beck, "Issues in the Use of an Electronic Rehabilitation System with Chronic Recidivists," *Law and Society Review* 3, no. 4 (May 1, 1969): 597–615.

176 "museums or monuments": Ibid.

177 "Is there also a broad hint": Jessica Mitford, *Kind and Usual Punishment* (New York: Vintage, 1974), 237.

178 The decision established: On reasonable expectations of privacy, see Zoila Hinson, "GPS Monitoring and Constitutional Rights," *Harvard Civil Rights–Civil Lberties Law Review* 43 (2008): 285–7; John S. Ganz, "It's Already Public: Why Federal Officers Should Not Need Warrants to Use GPS Vehicle Tracking Devices," *Journal of Criminal Law and Criminology* 95, no. 4 (Summer 2005): 1325–62.

182 Satellite dishes on the trucks: Agis Salpukas, "Business Technology; Satellite System Helps Trucks Stay in Touch," *New York Times*, June 5, 1991.

183 1.3 million fleet vehicles: Ganz, "It's Already Public."

183 More than half: Aaron Huff, "Market for GPS Fleet Management Still Poised for Growth," *Commercial Carrier Journal*, February 3, 2014.

183 The worldwide fleet management industry: "Fleet Management Market by Components, Technologies and Services (Fleet Analytics, Vehicle Tracking and Fleet Monitoring, Telemetric, Vendor Services), by Fleet Vehicle Types (Trucks, Light Goods, Buses, Corporate Fleets, Container Ships, Aircrafts)—Global Forecast to 2019," MarketsandMarkets, 2015.

183 In Delhi: Manju Bansal, "India's GPS-enabled Rickshaws," SAP News Center, July 17, 2013, http://news.sap.com/indias-gps-enabled-rickshaws/.

183 In China: Eileen Yu, "Chinese City Uses GPS to Address Food Waste," ZDNet, August 6, 2013, http://www.zdnet.com/article/chinese-city-uses-gps-to-address-food-waste/.

184 "As soon as the employees . . . driver time savings": statistics and testimonials from "Survey of Fleet Operator Interest in MRM Systems and Services" (Palos Verdes Estates, CA: C. J. Driscoll and Associates, October 2013).

186 the German decision: On the legal significance of this case, see Jacqueline E. Ross, "Germany's Federal Constitutional Court and the Regulation of GPS Surveillance," *German Law Journal* 6, no. 12 (2005): 1805–12.

187 The suspect had also argued: On the decision by the European Court of Human Rights, see Bernadette Rainey, Elizabeth Wicks, and Clare Ovey, *Jacobs, White, and Ovey: The European Convention on Human Rights* (Oxford: Oxford University Press, 2014), 374.

187 In the U.K., the law: On the classifications of surveillance in the U.K. and

Ireland and their relation to GPS tracking, see Liz Campbell, *Organised Crime and the Law: A Comparative Analysis* (Oxford: Hart, 2013), 77–81; Liz Campbell, "GPS Surveillance: US v European Jurisdictions," *IntLawGrrls*, February 22, 2012, http://www.intlawgrrls.com/2012/02/gps-surveillance-us-v-european .html.

189 The police, government lawyers argued: My understanding of *United States v. Jones* was informed by an amicus brief in support of Jones submitted by the Center for Democracy and Technology and the Electronic Frontier Foundation, cosigned by GPS godfather Roger Easton, as well as by an interview with Jim Dempsey, vice president for public policy at the Center for Democracy and Technology.

190 "In effect": Lyle Denniston, "Opinion Recap: Tight Limit on Police GPS Use (Final Update)," SCOTUS blog, January 23, 2012, http://www.scotusblog .com/2012/01/opinion-recap-tight-limit-on-police-gps-use/.

191 "build situational contexts": Mary Shacklett, "GPS Serves a Role in Big Data Stickiness," *TechRepublic*, February 27, 2015, http://www.techrepublic.com/ article/gps-serves-a-pivotal-role-in-big-data-stickiness/.

192 "There is a heartbeat pattern": Celeste Arbogast, "Taxi GPS Data Helps Researchers Study Hurricane Sandy's Effect on NYC Traffic," University of Illinois at Urbana–Champaign College of Engineering, October 25, 2014, http://engineering.illinois.edu/news/article/9717.

195 first wave of these house arrest systems: Joseph Hoshen, George Drake, and Debra D. Spencer, "Wide-Area Continuous Offender Monitoring", *Proceedings of the International Society for Optics and Photonics* 2938 (1997), 166.

195 1987–94 statistics: J. Hoshen, J. Sennott, and M. Winkler, "Keeping Tabs on Criminals," *IEEE Spectrum* 32, no. 2 (February 1995): 26–32.

197 80,000 people each day: Estimate provided by Mike Nellis. Other information regarding GPS offender tracking is drawn from Mike Nellis, "The GPS Satellite Tracking of Sex Offenders in the USA," in *Sex Offenders: Punish, Help, Change or Control?*, edited by Jo Brayford, Francis Cowe, and John Deering (London: Routledge, 2012), 246–64.

197 The exception is Great Britain: On the recent history of offender monitoring in the U.K., see Mike Nellis, "Upgrading Electronic Monitoring, Downgrading Probation: Reconfiguring 'Offender Management' in England and Wales," *European Journal of Probation* 6, no. 2 (August 2014): 169–91.

201 His name was Gary Bojczak: Details of the apprehension of Bojczak are largely drawn from the FCC's "Notice of Apparent Liability For Forfeiture," which identifies the agency's target on August 4, 2012, as Gary P. Bojczak of Whitehouse Station, NJ. Several media outlets reported that Bojczak worked for a construction firm, based on a LinkedIn profile of a Gary Bojczak of Whitehouse Station, NJ, which listed Bojczak as "Lead Engineer at Tilcon New York Inc." From what I can tell, no reporters were able to contact Bojczak for independent confirmation. My own attempts were unsuccessful. I was also unable

to get Tilcon to confirm that Bojczak was their employee. The fact that Tilcon received a contract to rehabilitate runway 44-22L, which lies close to the Turnpike, leads me to believe that Bojczak was likely a Tilcon employee, on the airport grounds in some sort of official capacity, when he was using his jammer. (See "Tilcon Manages Major Project at Newark Airport," http://tilconny.com/projectDetail.htm?Tilcon-Manages-Major-Project-at-Newark-Airport.)

Chapter 8: Return from Mid-Ice

202 a city the size of Los Angeles: Marcia McNutt, director of the U.S. Geological Survey, has made this comparison; see "Japan Quake Aftershocks Could Go 'For Years,'" CBS News, March 12, 2011, http://www.cbsnews.com/news/japan-quake-aftershocks-could-go-on-for-years/.

202 The last mega-quake: Hector Becerra, "Likelier Here: the Next Big One," *Los Angeles Times*, April 15, 2008.

203 more to fear today from the Hayward Fault: Yehuda Bock, author interview.

204 Others who noted: Wolf Uwe Reimold, "Revolutions in the Earth Sciences: Continental Drift, Impact and Other Catastrophes," *South African Journal of Geology* 110, no. 1 (2007): 1–46, http://sajg.geoscienceworld.org/content/110/1/1.short.

205 "He could see more profoundly": Johannes Georgi, *Mid-Ice: The Story of the Wegener Expedition to Greenland*, translated by F. H. Lyon (London: Kegan Paul, Trench, Trubner & Co., 1954), 12.

206 "a selective search," "fairy tale": Quotes from Berry and Willis are in Hal Hellman, *Great Feuds in Science: Ten of the Liveliest Disputes Ever* (New York: John Wiley & Sons, 1998), 149–150.

206 "not seeking truth": Roger M. McCoy, *Ending in Ice: The Revolutionary Idea and Tragic Expedition of Alfred Wegener* (Oxford: Oxford University Press, 2006), 33.

206 "If we are to believe": Richard Conniff, "When Continental Drift Was Considered Pseudoscience," *Smithsonian Magazine*, June 2012, http://www.smithsonianmag.com/science-nature/when-continental-drift-was-considered-pseudoscience-90353214/.

206 "a drying apple": Alfred Wegener, *The Origin of Continents and Oceans*, translated by J. G. A. Skerl (London: Methuen, 1924), 12.

206 "It is inconceivable": Willem Anton Josef Maria van Waterschoot Van der Gracht, ed., *Theory of Continental Drift: A Symposium on the Origin and Movement of Land Masses, Both Inter-Continental and Intra-Continental, as Proposed by Alfred Wegener* (Tulsa, OK: American Association of Petroleum Geologists, 1928), 195.

207 Mid-Atlantic Ridge: Hellman, *Great Feuds In Science*, 153.

207 geochronology technologies: Reimold, "Revolutions in the Earth Sciences," 9.

208 When a continental plate: For a concise early account of the emerging science

of plate tectonics, see Allen W. Levy, "Movers In the Earth," *The Sciences* 15, no. 9 (December 1975): 6–10.

208 "The planet earth": Massachusetts Institute of Technology, "The Terrestrial Environment: Solid-Earth and Ocean Physics," NASA Contractor Report 1579, April 1970, 7–14.

210 In 1978, hundreds of scientists: Steve Harvey, "Palmdale Bulge Was a Mountain of Mystery," *Los Angeles Times*, December 10, 2010.

212 Canberra presentation: Charles Counselman, author interviews, and Jill Hecht Maxwell, "Charles C. Counselman III '64, SM '65, PhD '69," *MIT Technology Review*, December 21, 2010, http://www.technologyreview.com/article/422110/charles-c-counselman-iii-64-sm-65-phd-69/.

213 "were consistent at a level": Larry D. Hothem and Charles J. Fronczek, "Report on Test and Demonstration of Macrometer Model V-1000 Interferometric Surveyor," Federal Geodetic Control Committee, Instrument Subcommittee, May 1983.

214 California Permanent GPS Geodetic Array: Y. Bock, S. Wdowinski, P. Fang, J. Zhang, S. Williams, H. Johnson, J. Behr, et al., "Southern California Permanent GPS Geodetic Array: Continuous Measurements of Regional Crustal Deformation between the 1992 Landers and 1994 Northridge Earthquakes," *Journal of Geophysical Research: Solid Earth* 102, no. B8 (August 10, 1997): 18013–33.

220 Verrazano-Narrows suspension bridge: Mikhail G. Kogan, Won-Young Kim, Yehuda Bock, and Andrew W. Smyth, "Load Response on a Large Suspension Bridge during the NYC Marathon Revealed by GPS and Accelerometers," *Seismological Research Letters* 79, no. 1 (2008): 12–19.

221 On the afternoon of March 11: Joe Burgess, Jonathan Corum, Amanda Cox, Matthew Ericson, G.V. Xaquín, Alan McLean, Tomoeh Murakami-Tse, et al., "How Shifting Plates Caused the Earthquake and Tsunami in Japan," *New York Times*, March 13, 2011.

222 strong ground-shaking: Diego Melgar, Brendan W. Crowell, Yehuda Bock, and Jennifer S. Haase, "Rapid Modeling of the 2011 Mw 9.0 Tohoku-Oki Earthquake with Seismogeodesy," *Geophysical Research Letters* 40, no. 12 (June 28, 2013): 2963–8.

225 Using the combined real-time data: Ibid.

225 the underestimated magnitude: Nam Yi Yun and Masanori Hamada, "Evacuation Behavior and Fatality Rate during the 2011 Tohoku-Oki Earthquake and Tsunami," *Earthquake Spectra* 31, no. 3 (May 1, 2014): 1237–65.

226 "We did this big earthquake scenario": "The 2008 Great Southern California ShakeOut Scenario," ShakeOut, http://www.shakeout.org/california/scenario/. Accessed November 2, 2015.

228 a method of using these bounces: Clara C. Chew, Eric E. Small, Kristine M. Larson, and Valery U. Zavorotny, "Effects of Near-Surface Soil Moisture on GPS SNR Data: Development of a Retrieval Algorithm for Soil Moisture,"

IEEE Transactions on Geoscience and Remote Sensing 52, no. 1 (January 2014): 537–43.

230 around $2 billion: Justin Bachman, "Iceland Sees a Potential Volcanic Eruption, and Airlines Cower," *Bloomberg Business*, August 18, 2014.

231 One of the most powerful tools: For a detailed description of GRACE, see Charles Dunn, Willy Bertiger, Garth Franklin, Ian Harris, Gerhard Kruizinga, Tom Meehan, Sumita Nandi, et al., "The Instrument on NASA's GRACE Mission: Augmentation of GPS to Achieve Unprecedented Gravity Field Measurements," *Proceedings of the Fifteenth International Technical Meeting of the Satellite Division of the Institute of Navigation*, Portland, OR, 2002 (ION GPS 2002), 724–30.

232 During 2010, some of the stations rose: Michael Bevis, John Wahr, Shfaqat A. Khan, Finn Bo Madsen, Abel Brown, Michael Willis, Eric Kendrick, et al., "Bedrock Displacements in Greenland Manifest Ice Mass Variations, Climate Cycles and Climate Change," *Proceedings of the National Academy of Sciences* 109, no. 30 (2012): 11944–8; "Warm Spike in 2010 Caused Greenland to Rise," *Livescience.com*, December 13, 2011, http://www.livescience.com/17457-2010-warm-spike-caused-greenland-rise.html.

233 "could hardly get over this": Georgi, *Mid-Ice*, 118.

234 According to some calculations: McCoy, *Ending In Ice*, 142–3.

Chapter 9: Tied Together

239 "GPS receivers without GIS": Kumar and Moore, "The Evolution of Global Positioning System Technology," 67.

239 An MIT Media Lab study: Philip Salesses, Katja Schechtner, and César A. Hidalgo, "The Collaborative Image of the City: Mapping the Inequality of Urban Perception," *PLoS ONE* 8, no. 7 (July 24, 2013).

241 measurable environmental impact: Daniel Strebe, "Why Mercator For the Web? Isn't Mercator Bad?" Mapthematics Forums, March 15, 2012, https://www.mapthematics.com/forums/viewtopic.php?f=8&t=251.

242 "buy, take out of box": Juliette Garside, "Harold Goddijn: TomTom's Founder Needs His Business to Turn the Corner," *Guardian*, November 24, 2011.

245 The related practice of land surveying: On early geodesy and surveying, see Irene K. Fischer, "At the Dawn of Geodesy," *Bulletin Géodésique* 55, no. 2 (1981): 132–42.

249 By the start of World War II: Richard K. Burkhard, "Geodesy For the Layman," 5th edition, U.S. Department of Commerce/National Oceanic and Atmospheric Administration/National Ocean Service, 1983.

251 preparing to attack the Marshalls: John Dille, "The Missile-Era Race to Chart the Earth," *Life*, May 12, 1958.

252 "This is not so silly": D. C. Martin, "The International Geophysical Year," *Geographical Journal* 124, no. 1 (1958): 18–29.

252 "For all our talk": Lang, "Earth Satellite No. 1."

252 "The United States": "A Scientific Satellite Program," Naval Research Laboratory Memorandum Report no. 487, July 5, 1955.

253 the most underappreciated scientific achievement: The historian John Cloud has called the WGS "arguably one of the most important American intellectual achievements of the Cold War . . . the culmination of an international intellectual endeavor that spanned at least two centuries": John Cloud, "American Cartographic Transformations during the Cold War," *Cartography and Geographic Information Science* 29, no. 3 (2002): 261–82.

254 official maps in Burkina Faso: Alain Bagre, Moha El-ayachi, and Ahomaki Tapio, "Establishing a Land Policy Reform and GPS Technology Implementation in Burkina Faso." World Bank, Washington, DC, April 8–11, 2013.

256 By controlling WGS 84: Barbara Wiley, David Craig, Dennis Manning, John Novak, Randall Taylor, and Leonard Weingarth, "NGA's Role in GPS," *Proceedings of the Nineteenth International Technical Meeting of the Satellite Division of the Institute of Navigation*, Fort Worth, TX, 2006 (ION GNSS 2006), 2111–19.

256 monitoring stations around the world: Brent Renfro, David Munton, Richard Mach, and Randall Taylor, "Around the World for 26 Years—A Brief History of the NGA Monitor Station Network," *Proceedings of the Institute of Navigation 2012 International Technical Meeting*, Newport Beach, CA, 2012, 1818–32.

260 Delta-DOR: David W. Curkendall and James S. Border, "Delta-DOR: The One-Nanoradian Navigation Measurement System of the Deep Space Network—History, Architecture and Componentry," *Interplanetary Network Progress Report* 42 (2013): 193; Roberto Maddè, Trevor Morley, Ricard Abelló, Marco Lanucara, Mattia Mercolino, Gunter Sessler, and Javier de Vicente, "Delta-DOR: A New Technique For ESA's Deep Space Exploration," *ESA Bulletin* 128 (November 2006): 68–74.

Epilogue: Direction Home

263 otherwise known as Tupaia's map: Anne Di Piazza and Erik Pearthree, "A New Reading of Tupaia's Chart," *Journal of the Polynesian Society* 116, no. 3 (2007): 321–40.

264 contacts had been severed for at least 500 years: Ben Finney, *Hokule'a: The Way to Tahiti* (New York: Dodd, Mead & Co., 1979), 9–10.

265 Navigating an inter-atoll voyage: Ibid., 125.

266 "I cannot see the island": Nainoa Thompson, "Reflections on Mau," Hawaiian Voyaging Traditions, http://pvs.kcc.hawaii.edu/index/founder_and_teachers/mau.html.

Acknowledgments

To employ a wincingly obvious analogy, researching and writing this book often felt like navigating a barren desert on a moonless night, without GPS. I was very fortunate to find beacons that helped me find my way.

In a publishing world struggling to find its own bearings, I assumed editors like Tom Mayer no longer existed. Tom helped me realize what this book is about. He understood, often before I did, how disparate historical threads could be wound into a thematic rope. His extraordinary focus, attention to detail, and insistence on drum-tight text brought the book into full relief. In the alternate quantum universe where someone else edited *Pinpoint*, it is a vastly inferior work.

If there is a publisher more supportive of its authors than Norton, I haven't heard of it. Ryan Harrington provided crucial editorial input and assistance. Ingsu Liu and her team orchestrated the striking cover art. Bill Rusin helped get the book into stores (and pixels), and Alice Rha got the word out. Laura Goldin ensured the coordinates were correct. Nancy Palmquist held it together. Allegra Huston let nothing get by her.

Bella Lacey read the manuscript, pointing out crucial questions that were going unanswered and providing a forest-level perspective when I was fixated on trees. She offered brilliant suggestions for restructuring as well as a much-needed reminder that although GPS has an American pedigree, its story needed to be global in scope.

As always, Daniel Greenberg was there from the fledgling concept to the finish line. I couldn't ask for a better advocate.

I was extremely fortunate to meet Len Jacobson, who spent hours giving me an insider's perspective on GPS history, and also gave the manuscript a full read. Read Len's memoir, *Flying For GPS*. Also, his book *GNSS Markets and Applications* is the best GPS primer I've come across.

Yehuda Bock generously gave an early version of chapter 8 a close read. Tom Herring and Kristine Larson also offered very useful critiques. Todd Humphreys and Roger Johnston weighed in on the spoofing material, as did Mike Nellis on GPS tracking.

Many people contributed to making GPS a reality—and I regret that this book inevitably leaves many unnamed—but two in particular deserve credit for creating the world's only global utility. Brad Parkinson had the vision and the drive to lead the team that created a technology that thrives today, having barely changed since the early 1970s. I'm very grateful for the large blocks of time he set aside to speak with me. I regret that I was never able to speak to Roger Easton, who died in 2014. Fortunately for me, his son, Richard Easton, has worked tirelessly to preserve his father's legacy—and helped me describe Roger's contributions to GPS history. Richard's book, *GPS Declassified*, is definitely worth a read. Leo Slater at the Naval Research Laboratory was also a wellspring of information.

Charlie Trimble, the "Brad Parkinson" of commercial GPS, also shared his story, and helped me untangle technological issues related to satellite navigation. Thanks also to the other original GPS visionaries who told their tales, Javad Ashjaee, Ralph Eschenbach, and Ed Tuck.

Very special thanks to Rita Chretien for sharing her truly amazing survival story.

Jennifer Thibault and the U.S. Air Force Book Support Program laid the groundwork for my visit to the GPS Master Control Station. At Schriever Air Force Base, Marie Denson made the necessary arrangements and was a gracious host, even though I got lost on the way and showed up late. (I blame GPS.) Stephen Dirks, Dean Holthouse, Jason Hope, Kevin Kiser, and Brian Stewart gave me the lowdown. Many thanks to Air Force Space Command, the 50th Space Wing, and especially the men and women of 2 SOPS, the tip of the GPS spear.

For their time and hospitality at NASA's Jet Propulsion Laboratory, I thank Yoaz Bar-Sever and Larry Young. JPL's Stephen Lichten and Tomas Martin-Mur were also invaluable sources of information on the intricacies of Mars landings.

Carol Snyder and Wade Stewart at Trimble's agriculture division went above and beyond in giving me an introduction to precision agriculture, as did Troy Seaworth and Seaworth Farms. Thank you to Silvia McLachlan for helping me make the connections.

Keith Davio arranged for me to tag along with one of TomTom's data collection vans—and thank you to Joe Palatucci for showing me a day in a driver's life. Dan Adams, Peter Davie, and Maureen Krauland also helped me navigate through the company.

Thank you to Tyler Barnet and Glenn Marston for responding to my quixotic attempts to solve the mystery of the Fallen Man on East 29th St.—and to Sara Murray of Buddi, for letting me get the full experience of being GPS-tracked, with the assistance of Louise Harrold and Chris Kennedy.

Alissa Kleinman provided impeccable research assistance.

Many thanks to everyone else who unhesitatingly extended a helping hand for this book, taking the time to tell me their stories, explaining the history and technological intricacies of GPS: Claudio Aporta, Michael Barclay, Beth Bartel, Ben Bartolome, Ron Beard, Eric Beaton, Michael Bevis, Steve Bradford, Justin Brookman, Nolan Bushnell, Robert Cheetham, Ann Ciganer,

Hervé Clauss, Steve Coast, Dan Cole, Peggy Conway, Chuck Counselman, Jim Davis, Loren De Groot, Jim Dempsey, Deborah Dennard, George Drake, Ryan Driscoll, Per Enge, Ralph Eschenbach, John Fischer, Sean Fitzpatrick, Julia Frankenstein, Gary Freeland, Robert Gable, Nunzio Gambale, Amy Gilroy, Buster Glosson, Allen Goldstein, Gaylord Green, Liz Groff, Anurag Gupta, Tim Hall, Rick Hamilton, Debbie Henderson, Jim Higgins, Kirk Holub, Chuck Horner, Joseph Hoshen, Jiung-Yao Huang, Ken Hudnut, Aaron Huff, Eric Hunsader, Michael E. Jackson, Kevin Kelly, Josh Khani, Jay Dee Krull, Gilly Leshed, Judah Levine, Sam Liang, Rich Maher, Mani Mahjouri, Steve Malys, Hans Mark, Susan Marshall, Tom McHugh, Jules McNeff, Donald Mitchell, Richard Nimer, Scott Pace, Dave Paddock, Ganesh Pattabiraman, Stig Pedersen, Bruce Peetz, Arun Phadke, Sam Pullen, Logan Scott, Edwin Stear, Sam Stein, Ron Stewart, Mike Swiek, Dave Van Dusseldorp, John Warburton, Jim White, Pete Wilhelm, Ronald Yates.

These people all made the task at hand a little easier: Peter Adams, Berni Ai-Kuo, Monica Allen, Holly Baker, Gary Barta, Beth Bartel, Jennifer Beadling, Philip Bourekas, Tim Boyle, James Burgess, Susan Cadrecha, John Ciampa, Steve Coast, Robert Crane, Cam Crary, William Davenport, Ann Marie Dryden, Douglas Equils, Erica Fouché, Johnine Fregia, Scott Gastel, Chris Gates, Malcolm Gladwell, Nancy Greco, Timothy Greeley, Ivan Isakova, LaGina Jackson, Jake Jacobson, Tammy Jones, Reishia Kelsey, Osama Khalil, Aaron Lebovitz, Rita Lee, Michael Lombardi, Peter Lumsdaine, Tom Mackie, Rod MacLeod, Michael Mathog, Donna McKinney, John Merrill, Tim Midyett, Kimberly Mielcarek, Manoj Narang, Frank O'Donnell, Kevin Ortiz, Emmanuelle Parise-Tarquis, Wyn Partington, Brian Peterson, Steven Romalewski, Dan Roman, Tracy Schriver, Brian Shaw, Paula Shawa, Paul Smith, Tom Soler, Kurt Steinert, Lonnie Tebow, John Tipaldo, Jack Tisdale, Jim Tugman, Rex Turner,

Pete VanAmburgh, Michael Wallace, Patrick Whalen, Ken White, Vladimir Yakovlev.

Thank you also to Brooklyn Public Library, Esri, FAA's William J. Hughes Technical Center, Granta Books, Metropolitan Transportation Authority (New York), National Geospatial-Intelligence Agency, New York State Department of Transportation, New York Public Library's Science, Industry and Business outpost, Stanford GPS Laboratory, Tradeworx, Transportation Alternatives, UNAVCO, Vulnerability Assessment Team.

Finally, my personal reference islands—even when I can't see them, I know they're there. My parents, Neal and Joy Milner, and my sister, Joanna Milner, have been unerringly supportive. Julie Taraska is the rock, the co-conspirator, the brightest beacon. Vivienne Milner was born around the time I began work on this book. It's been a privilege watching her cognitive map grow more comprehensive at the same time as I was piecing this all together. And now the playground awaits.

Index

Page numbers in *italics* refer to illustrations. Page numbers beginning with 267 refer to endnotes.

OCEAN COUNTY LIBRARY